No. 256

# Essential Cell Biology

## Volume 2: Cell Function

## A Practical Approach

Edited by

John Davey and Mike Lord

Department of Biological Sciences,
The University of Warwick,
Coventry CV4 7AL, UK

OXFORD
UNIVERSITY PRESS

Great Clarendon Street, Oxford OX2 6DP

Oxford University Press is a department of the University of Oxford.
It furthers the University's objective of excellence in research, scholarship,
and education by publishing worldwide in

Oxford New York

Auckland Bangkok Buenos Aires Cape Town Chennai Dar es Salaam
Delhi Hong Kong Istanbul Karachi Kolkata Kuala Lumpur Madrid
Melbourne Mexico City Mumbai Nairobi São Paulo Shangai Taipei
Tokyo Toronto

Oxford is a registered trade mark of Oxford University Press in the UK and
in certain other countries

Published in the United States by Oxford University Press Inc., New York



First published 2003



A catalogue record for this title is available from
the British Library

Library of Congress Cataloguing in Publication Data
Essential cell biology : a practical approach / edited by John Davey
and Mike Lord.
p. cm. (Practical approach series ; 262–)
Includes bibliographical references and index.
Contents: v. Cell structure – v. 2. Cell function.
1. Cytology–Laboratory manuals. I. Davey, John. II. Lord, Mike.
III. Practical approach series ; 262, etc.
QH583.2. E85 2002 571.6–dc21 2002029013

ISBN 0 19 963830 6 (v. 1 : hbk.) ISBN 0 19 963832 2 (v. 2 : hbk.)
ISBN 0 19 963831 4 (v. 1 : pbk.) ISBN 0 19 963833 0 (v. 2 : pbk.)
10 9 8 7 6 5 4 3 2 1

Typeset in Swift by Footnote Graphics Ltd, Warminster, Wilts
Printed in Great Britain on acid-free paper
by The Bath Press Ltd, Avon

# Preface

Cell biology relies upon an integrated understanding of how molecules within cells interact to carry out and regulate the processes required for life. Obtaining such an integrated understanding imposes great demands on today's investigators. Being an expert in a relatively narrow area is no longer sufficient and researchers need to be able to call upon a battery of techniques in their quest for information. Unfortunately, a lack of familiarity with the experimental possibilities can often discourage diversification. This two-volume set is designed to try to help overcome some of these problems. With the help of experienced researchers, we have been able to gather together a compendium of protocols that covers most of the approaches available for studying cell biology. Inevitably in a project of this size it has not been possible to include every technique but with Volume 1 focusing on the techniques for studying cell structure and Volume 2 concentrating on understanding how the cell functions, we believe that all of the essential protocols are included. Both traditional and more modern techniques are covered and the theory and principles of each are described, together with detailed protocols and advice on trouble shooting. Directions to more specialized developments are also included. Although written by experts, each section is accessible to those new to science. We hope the result inspires readers to experience the challenges and rewards of entering a new area of cell biology for themselves.

We thank all of the authors for agreeing to undertake the time-consuming job of preparing their chapters, and the staff at Oxford University Press for their help at all stages of preparation and production. Finally, we would appreciate receiving any comments on the text and the correction of any errors that might have been missed.

Warwick J. D. and J. M. L.
November 2002

# Contents

# Protocol list

# Abbreviations

| | |
|---|---|
| AEBSF | aminoethyl-benzenesulfonyl fluoride |
| BHK | baby hamster kidney |
| bHRP | biotinylated HRP |
| BrUTP | 5-bromouridine 5′-triphosphate |
| BSA | bovine serum albumin |
| cdk | cyclin-dependent kinase |
| cpm | counts per minute |
| CSK | cytoskeleton buffer |
| DAG | diacylglycerol |
| DAPI | 4′,6-diamidino-2-phenylindole dihydrochloride |
| DCF | dichlorofluorescein |
| DCF-$H_2$ | dichlorofluorescin |
| DEPC | diethyl pyrocarbonate |
| $\Delta p$ | protonmotive force |
| $\Delta pH$ | pH gradient across the inner mitochondrial membrane |
| $\Delta\psi_m$ | mitochondrial membrane potential |
| DMF | dimethylformamide |
| DMSO | dimethyl sulfoxide |
| DOP-PCR | degenerate oligonucleotide-primed polymerase chain reaction |
| DTT | dithiothreitol |
| ECV/MVB | endosomal carrier vesicles/multivesicular bodies |
| ER | endoplasmic reticulum |
| FCCP | carbonylcyanide-*p*-trifluoromethoxyphenyl hydrazone |
| FISH | fluorescence *in situ* hybridization |
| FRET | fluorescence resonance energy transfer |
| FSG | fish skin gelatin |
| GFP | green fluorescent protein |
| GST | glutathione *S*-transferase |
| HA | haemagglutinin |
| hnRNPs | heterogeneous nuclear ribonucleoproteins |
| HRP | horseradish peroxidase |
| IF | indirect immunofluorescence |

| | |
|---|---|
| ISH | *in situ* hybridization |
| LOH | loss of heterozygosity |
| MEF | mouse embryonic fibroblast |
| NLS | nuclear localization signal |
| PBS | phosphate-buffered saline |
| PEI | polyethylenimine |
| PI | propidium iodide |
| PKC | protein kinase C |
| PLC | phospholipase C |
| PLD | phospholipase D |
| PMSF | phenylmethylsulfonyl fluoride |
| RNA pol | RNA polymerase |
| SMG | small molecular weight GTPases |
| TCA | trichloroacetic acid |
| $TMRM^+$ | tetramethylrhodamine methyl ester |
| tRNA | transfer RNA |

# Chapter 1
# Enzyme activities

Keith F. Tipton
Department of Biochemistry, Trinity College, Dublin 2, Ireland.

## 1 Introduction

This account will concentrate on the behaviour of enzymes, the determination of their activities, and their responses to alterations in substrate concentration. Particular emphasis will be placed on studies of the effects of inhibitors, since these may be used as therapeutic drugs as well as tools to help elucidate their mechanisms of action and behaviour in the cell.

## 2 Enzyme specificity and nomenclature

Enzymes are highly specific catalysts that each use one or a relatively small group of related compounds as substrates. As with all aspects of biology it is necessary to have some defined classification system to minimize ambiguities; in the past there have been many instances of people using the same name for different enzymes. Unlike protein and gene nomenclature systems, enzymes are classified according to the reactions that they catalyse. The IUBMB Enzyme Nomenclature web site (1) contains a listing of all classified enzymes. Each enzyme is allocated a four-digit EC number, the first three of the digits define the reaction catalysed and the fourth digit is a unique identifier. The reaction catalysed is given and there is a systematic name that uniquely defines the reaction catalysed. A simpler recommended name, usually one that is in common use is assigned and other names that have been used are also listed (see ref. 2 for fuller details of the enzyme classification and nomenclature system). Use of the recommended names and EC number in publications eliminates possible ambiguities and facilitates the searching of literature databases. That web site (1) also links each enzyme to other databases, including those on genes and protein sequences, enzyme properties, and metabolic involvement.

## 3 Determination of enzyme activity

The activity of an enzyme may be measured by determining the rate of product formation or substrate utilization during the enzyme-catalysed reaction. For many enzymes there are several alternative assay procedures available and the choice between them may be made on the grounds of convenience, cost, the availability of appropriate equipment and reagents, and the level of sensitivity required. Convenient recipes for the assay of many individual enzymes can be found in a variety of sources, such as *Methods in enzymology* (3), *Methods in enzymatic analysis* (4), and *The enzyme handbook* (5), as well as in the original literature, some of which can be conveniently assessed by way of the enzyme nomenclature web site (1).

## 3.1 Reaction progress curves

The time-course of product formation, or substrate utilization is often curved, as shown in Figure 1. It is initially linear but the rate starts to decline at longer times. This fall-off could result from one or more causes (see ref. 6 for more detailed discussion), including substrate depletion, approach to equilibrium of a reversible reaction, inhibition by the product(s) of the reaction, instability of the enzyme, time-dependent inhibition of the enzyme by the substrate, a change in assay conditions as the reaction progresses, or an artefact in the assay procedure. At very short times none of these effects should be significant and so if the initial, linear, rate of the reaction is determined by drawing a tangent to the early part of the progress curve (see Figure 1), these complexities should be avoided. The linear portion of an assay is usually long enough to allow the initial rate to be estimated accurately simply by drawing a tangent to, or taking the first-derivative of, the early part of the progress curve. It has sometimes been assumed that restricting measurements of reaction rates to a period in which less than 10–20% of the total substrate consumption has occurred will give an accurate measure of the initial rate. Consideration of the possible causes for non-linearity listed above (see also ref. 6) will show that such an approach may not be valid.

In cases where curvature makes it difficult to estimate the initial rate with accuracy, it may be possible to do so by fitting the observed time-dependence of

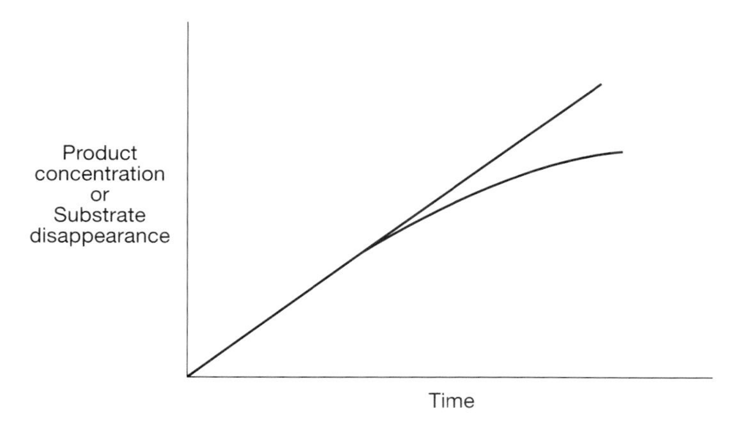


**Figure 1** A possible progress curve of an enzyme-catalysed reaction. The tangent to the initial, apparently linear, portion of the time-course indicates the initial rate of the reaction.

product formation to a polynomial equation and deriving the initial slope at t = 0 (see ref. 7). Graphs of [product]/time against either time or [product] will intersect the vertical axis at a point corresponding to the initial rate.

## 3.2 The effects of enzyme concentration and expression of activities

Enzymes are catalysts that are present at much lower concentrations than the substrate(s). Therefore the initial velocity of the reaction would be expected to be proportional to the concentration of the enzyme and this is the case for most enzyme-catalysed reactions, where a graph of initial velocity against total enzyme concentration gives a straight line passing through the origin (zero activity at zero enzyme concentration). Under these conditions, the ratio (velocity/enzyme concentration) will be a constant that can be used to express the activity of an enzyme quantitatively. This can be valuable for comparing data obtained with the same enzyme from different laboratories, assessing the effects of physiological or pharmacological challenges to cells or tissues, monitoring the extent of purification of enzymes, and comparing the activities of different enzymes, or of the same enzyme from different sources or with different substrates.

The activity of an enzyme may be expressed in any convenient units, such as absorbance change per unit time per mg enzyme protein, if a spectrophotometric assay is used, but it is preferable to have a more standardized unit in order to facilitate comparisons. The most commonly used quantity is the **Unit**, sometimes referred to as the International Unit or Enzyme Unit. One Unit of enzyme activity is defined as that catalysing the conversion of 1 μmol substrate (or the formation of 1 μmol product) in 1 min. The specific activity of an enzyme preparation is the number of Units per mg protein. Since the term 'unit' has also sometimes been used to refer to more arbitrary measurements of enzyme activity, it is essential that it is defined in any publication.

If the relative molecular mass of an enzyme is known it is possible to express the activity as the **molecular activity**, defined as the number of Units per μmol of enzyme; in other words the number of mol of product formed, or substrate used, per mol enzyme per min. This may not correspond to the number of mol substrate converted per enzyme active site per minute since an enzyme molecule may contain more than one active site. If the number of active sites per mol is known the activity may be expressed as the **catalytic centre activity**, which corresponds to mol substrate used, or product formed, per min per catalytic centre (active site). The term turnover number has also been used quite frequently but there appears to be no clear agreement in the literature as to whether this refers to the molecular or the catalytic centre activity.

Although the Unit of enzyme activity, and the quantities derived from it, are most widely used, the **Katal** (abbreviated to kat) is an alternative. This differs from the units described above in that the second, rather than the minute, is used as the unit of time in conformity with the International System of units (SI

Units). One Katal corresponds to the conversion of 1 mol of substrate per second. Thus it is an inconveniently large quantity compared to the Unit. The relationships between Katals and Units are:

$$1 \text{ kat} = 60 \text{ mol.min}^{-1} = 6 \times 10^7 \text{ Units}$$
$$1 \text{ Unit} = 1 \ \mu\text{mol.min}^{-1} = 16.67 \text{ nkat}$$

In terms of molecular, or catalytic centre, activities the Katal is, however, not such an inconveniently large quantity and it is consistent with the general expression of rate constants in $s^{-1}$.

When expressing the activity of an enzyme it is important to consider the stoichiometry of the reaction. Some enzyme-catalysed reactions involve two molecules of the same substrate. For example, adenylate kinase (EC 2.7.4.3) catalyses the reaction:

$$2\text{ADP} \rightleftharpoons \text{AMP} + \text{ATP}$$

Thus the activity will be twice as large if it is expressed in terms of ADP utilization than if expressed in terms of the formation of either of the products. Similarly, carbamoylphosphate synthetase (ammonia) (EC 6.4.3.16) catalyses:

$$\text{HCO}_3^- + 2\text{ATP} + \text{NH}_4^+ \longrightarrow \text{Carbamoylphosphate} + 2\text{ADP} + \text{P}_i$$

and the value expressed in terms of disappearance of ATP or formation of ADP would, thus, be twice that obtained if any of the other substrates or products were measured. Because of this, it is important to specify the substrate or product measured and the stoichiometry when expressing the specific activity of an enzyme.

There are, however, some cases where a simple proportionality between initial velocity and enzyme concentration does not appear to hold (see ref. 6) and it is, thus, always essential to check for linearity. In some cases departure from linearity may be artefactual resulting from, for example, failure to measure the initial rates of the reaction, the detection method becoming rate-limiting at higher enzyme concentrations, or from changes of the pH or ionic strength of the assay mixture as increasing amounts of the enzyme solution are added, and it is important to check that such effects are not occurring. In other cases the behaviour can be more interesting and a graph of initial velocity against enzyme concentration can show either upward or downward curvature. Upward curvature may result from the presence of a small amount of an irreversible inhibitor of the enzyme in the assay mixture or from the presence of a dissociable activator in the enzyme solution, whereas downward curvature may result from the presence of a dissociable inhibitor in the enzyme solution (see ref. 6).

## 3.3 Conditions for activity measurements

Although the quantity velocity/enzyme concentration is a useful constant for comparative purposes, it will only be constant under defined conditions of pH, temperature, and substrate concentration. A temperature of 30 °C is quite commonly used as a standard for comparative purposes and this is certainly more meaningful than 'room temperature', but in many cases it may be desirable to use a more physiological temperature. There is no clear recommendation as to pH and substrate concentration except that these should be stated and, where practical, should be optimal. However, it would be more appropriate to use physiological pH values, which may differ from the optimum pH for the reaction, if the results are to be related to the behaviour of the enzyme *in vivo*. Since the activities of some enzymes are profoundly affected by the buffer used and the ionic strength of the assay mixture, the full composition of the assay should be specified.

# 4 The effects of substrate concentration

## 4.1 The Michaelis–Menten relationship

For a simple enzyme-catalysed reaction:

$$E + S \underset{k_{-1}}{\overset{k_{+1}}{\rightleftharpoons}} E.S \xrightarrow{k_{cat}} E + \text{Products} \qquad [1]$$

the dependence of the initial rate ($v$) of product formation (or substrate disappearance) on the substrate concentration is given by the Michaelis-Menten relationship, which predicts a hyperbolic relationship between initial velocity and substrate concentration ($\boldsymbol{s}$) (see Figures 2 and 5):

$$v = \frac{k_{cat}\boldsymbol{e}.\boldsymbol{s}.}{K_m + \boldsymbol{s}} = \frac{V.\boldsymbol{s}}{K_m + \boldsymbol{s}} \qquad [2]$$

where $\boldsymbol{e}$ is the enzyme concentration, $V$ is the limiting velocity ($= k_{cat}\boldsymbol{e}$), and $K_m$, the substrate concentration necessary for the initial velocity ($v$) to equal $V/2$, is $(k_{-1} + k_{cat}) / k_{+1}$. The kinetic behaviour of many enzymes is described by this relationship. The kinetic parameters can be determined by directly fitting initial rate data to a rectangular hyperbola and there are several readily-available computer programs for doing this, e.g. ENZFITTER and ULTRAFIT (Elsevier Biosoft, Amsterdam), or MacCurveFit and WinCurveFit (8). Although the double-reciprocal plot is recognized to be an inaccurate procedure for determining enzyme kinetic parameters (see e.g. ref. 9), it is useful for illustrative purposes and will be used for such in this account.

## 4.2 Failure to obey the Michaelis–Menten equation

Departure from the simple hyperbolic behaviour predicted by the Michaelis–Menten equation can result from a number of different causes which have been discussed in detail by Tipton (6, 10). Assay artefacts, such as failure to determine the initial rate of the reaction, to subtract a blank-rate that occurs in the absence of all the reactants, to take account of the existence of a substrate–activator complex (see ref. 10 and Sections 5.1.7v and 5.1.8), or to take account of reaction

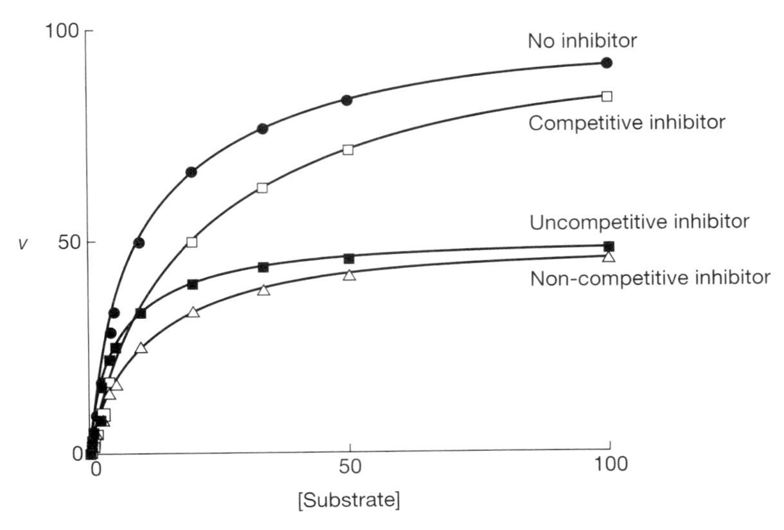


**Figure 2** The effects of substrate concentration on the rates of an enzyme-catalysed reaction in the presence of reversible inhibitors.

stoichiometry can lead to an apparently complex behaviour. Sigmoid dependence of velocity upon substrate concentration may indicate that the enzyme obeys cooperative kinetics. Cooperativity, as strictly defined, is a phenomenon reflecting the equilibrium binding of substrates, or other ligand, where the binding of one molecule of a substrate to an enzyme can either facilitate (positive cooperativity) or hinder (negative cooperativity) the binding of subsequent molecules of the same substrate (see Figure 3). In order to ensure that such behaviour does result from cooperativity, it would be necessary to perform substrate-binding studies under equilibrium conditions, since the steady-state initial velocities may not bear any simple relationship to the equilibrium saturation curve for substrate binding.

Most cooperative enzymes also exhibit allosteric behaviour. That is, their activities are affected by the binding of molecules (allosteric effectors) to sites distinct from the active site. Allosteric effects are, however, distinct from cooperativity and may occur in enzymes that show no cooperativity. Full discussions of the methods available for analysing cooperative behaviour and distinguishing between the possible models that may account for such effects are available elsewhere (see e.g. refs 11 and 12).

It is common to present data for cooperative enzyme in terms of the simple model advanced by Hill (13) in an attempt to explain the sigmoid saturation curve for oxygen binding to haemoglobin:

$$Ys = \frac{\boldsymbol{s}^h}{K + \boldsymbol{s}^h} \text{ or } v = \frac{V\boldsymbol{s}^h}{K + s^h} \qquad [3]$$

where $Y_s$ represents the fractional saturation of the enzyme with substrate and $h$ (also referred to as, among other things, n, $n_H$, or $n_h$) is the Hill constant, which

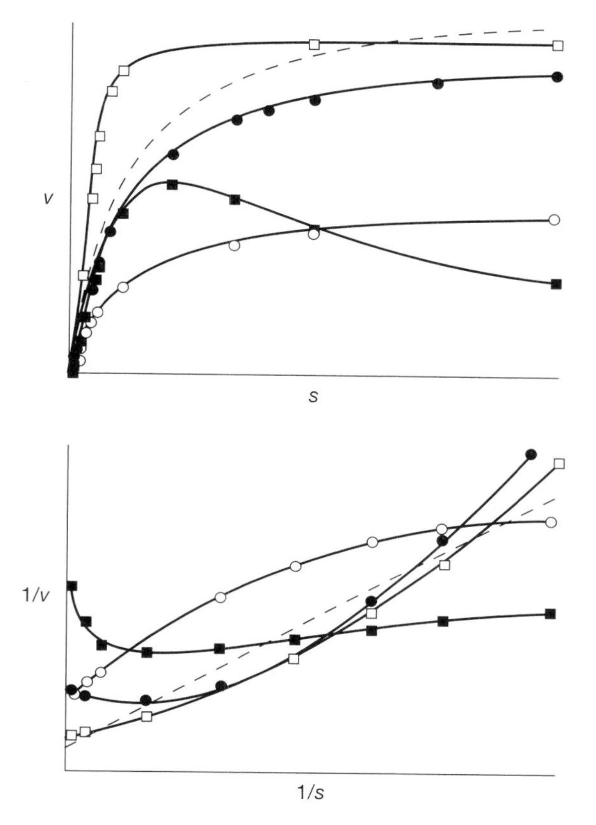


**Figure 3** Possible curves for the substrate concentration dependence for systems obeying Equation 7. The broken line represents simple Michaelis–Menten (Equation 2) behaviour.

does not necessarily correspond to the number of substrate-binding sites present in the molecule (see refs 11 and 12). The formulation for a cooperative enzyme where initial velocities are measured is also shown. The relationship can be transformed to a linear relationship as:

$$\log\left(\frac{v}{V - v}\right) = h\log \mathbf{s} - \log K \qquad [4]$$

Thus a graph of log $\{v / (V - v)\}$ against log **s** (a Hill plot) should give a straight line of slope $= -\log K$ and intercept on the $y$ axis $= h$. The Hill equation has been shown to be based on an inadequate model, because it envisages the simultaneous binding of all substrate molecules to the enzyme and it has been shown that for any system that involves the sequential binding of substrates, the plot will be linear only over a restricted range of substrate concentrations with the slopes tending to unity at very high and very low substrate concentrations (see refs 11 and 12). Nevertheless, the plot is still widely used to express cooperativity and the Hill constant is calculated from the region of maximum slope. A Hill constant of greater than unity indicates positive cooperativity whereas one of less than unity is given in cases of negative cooperativity. If there is no cooperativity the value of $h$ will be unity and Equation 4 will reduce to the simple Michaelis–Menten equation (Equation 2).

Any enzyme reaction mechanism in which there are alternative pathways by which the substrates can interact to give the complex that breaks down to give products will, under steady-state conditions, give rise to a complex initial rate equation. The simplest example concerns an enzyme with two substrates, Ax and B, which are converted to the products A and Bx, respectively. A reaction mechanism in which the two substrate can bind to the enzyme in a random order:

$$\mathrm{E} \rightleftarrows \begin{matrix} \mathrm{EAx} \\ \mathrm{EB} \end{matrix} \rightleftarrows \mathrm{EAxB} \dashrightarrow \mathrm{E{+}A{=}Bx} \qquad [5]$$

will give a steady-state initial rate equation of the form:

$$v = \frac{p(\boldsymbol{ax.b}) + q(\boldsymbol{ax^2b}) + r(\boldsymbol{ax.b^2})}{s + t(\boldsymbol{ax}) + u(\boldsymbol{b}) + v(\boldsymbol{ax.b}) + w(\boldsymbol{ax^2}) + x(\boldsymbol{b^2}) + y(\boldsymbol{ax^2b}) = z(\boldsymbol{ax.b^2})} \qquad [6]$$

where the constants $p$–$z$ are combinations of rate constants, and $\boldsymbol{ax}$ and $\boldsymbol{b}$ are the concentrations of Ax and B. At a fixed concentration of one of the substrates, for example B, this horrid equation may be simplified to:

$$\mathrm{v} = \frac{\alpha_1(\boldsymbol{ax}) + \alpha_2(\boldsymbol{ax^2})}{\beta_0 + \beta_1(\boldsymbol{ax}) + \beta_2(\boldsymbol{ax^2})} \qquad [7]$$

An equation of this form, containing squared terms in the substrate concentration, can give rise to a variety of curves describing the variation of initial velocity with substrate concentration including sigmoidal behaviour, as shown in Figure 3. The random-order mechanism will yield simple Michaelis–Menten behaviour if the rate of breakdown of the EAxB ternary complex to give products is slow relative to its rate of breakdown to the binary (EAx and EB) complexes so that the system remains in thermodynamic equilibrium (see refs 14–16).

Complexities of the steady-state reaction mechanism can result in quite weird kinetic behaviour with multiple inflection points or waves in the curves of initial velocity against substrate concentration (see ref. 17). True cooperativity in which an enzyme shows a mixture of positive and negative cooperativity for the successive binding of molecules of the same substrate can also give multiple inflection points and plateaus in the saturation, or velocity–substrate concentration curve (18). Complex negative cooperativity for multiple binding sites can result in curves of velocity against substrate concentration which may appear to be composed of linear sections with apparently sharp breaks between them (19, but see also ref. 20). Mechanisms involving enzyme isomerization can also give rise to complex initial rate behaviour.

A particular case which may be difficult to distinguish from negative cooperativity is where the same reaction is catalysed by two enzymes which have different $K_m$ values, the rate of the overall reaction will be given by:

$$v = \frac{V^a}{1 + K_m^a is} + \frac{V^b}{1 + K_m^b is} \qquad [8]$$

where $K_m{}^a$ and $V^a$ are the Michaelis constant and maximum velocity of one enzyme and $K_m{}^b$ and $V^b$ are the corresponding constants for the other. This equation will result in a behaviour similar to that of negative cooperativity. Thus it is always necessary to check for the presence of more than one enzyme if such behaviour is observed. This is particularly important since if the two enzymes had different binding affinities for their substrates, as well as $K_m$ values, similar behaviour would be shown in studies of the equilibrium binding of substrate. Equation 8 predicts a smooth curve when the data are plotted in double-reciprocal form (see Figure 3) without sharp breaks. It is not possible to obtain accurate estimates of the two $K_m$ and $V_{max}$ values by extrapolation of the apparently linear portions of the double-reciprocal plots at very high and very low substrate concentrations. Such an approach does not yield accurate values (see ref. 21) unless the difference between the $K_m$ values is greater than 1000-fold. It is possible, however, to determine the individual values using an iterative procedure that fits the data points to the sum of two rectangular hyperbolas. An easy manual way of doing this has been described (22) but curve-fitting using one of the many computer programs available (e.g. ref. 8) is straightforward. It is always good practice to construct a curve from the calculated values using Equation 8 and to compare it with the data points to ensure that a good fit has been obtained.

## 5 Enzyme inhibitors

Enzyme inhibitors are of value for understanding the behaviour of individual enzymes and metabolic systems. Furthermore, many drugs that are in clinical use work by inhibiting specific enzymes, either in the individual being treated or in an invading organism. An understanding of the different types of inhibition that can occur is necessary for understanding their likely effects in the tissues or organism and for the design of new drugs.

Enzyme inhibitors may interact with the target enzyme either reversibly or irreversibly. The behaviour of these two types is very different, both *in vivo* and *in vitro*. The basic differences between their behaviour are summarized in Table 1, where E is the enzyme, I the inhibitor, and E.I and E-I are non-covalently-bound and covalently-bound complexes between enzyme and inhibitor, respectively.

### 5.1 Reversible inhibitors

Reversible inhibition usually occurs extremely rapidly because there is no chemical reaction involved, there is simply a non-covalent interaction (Table 1). There are several types of reversible inhibitor which can be distinguished in terms of their effects with respect to substrate concentration. The different types of inhibitor may be distinguished by their effects on the Michaelis constant ($K_m$), the limiting (maximum) velocity ($V$), and the ratio $V/K_m$ of the enzyme-catalysed reaction.

**Table 1** Some basic differences between reversible and irreversible inhibitors

| Inhibition mechanism | Kinetic constant | Rate of inhibition | Reversibility *In vitro* | *In vivo* |
|---|---|---|---|---|
| **Reversible** | | | | |
| $E + I \underset{k_{-1}}{\overset{k_{+1}}{\rightleftharpoons}} E.I.$ | $K_i = \frac{k_{-1}}{k_{+1}}$ | Usually fast | Dialysis, dilution, gel filtration | Elimination of free inhibitor |
| **Irreversible** | | | | |
| $E + I \xrightarrow{k} E\text{–}I$ | $k$ | Often slower | None | Synthesis of more enzyme |

Inhibitor potency is usually expressed in terms of the dissociation constant ($K_i$) as described in Table 1. The use of dissociation constants remains an arcane feature of biochemistry (chemists generally prefer association constants) and means that $K_i$ is an inverse measure of enzyme inhibitor affinity; the lower the $K_i$ the higher the affinity.

### 5.1.1 Competitive inhibition

This can be represented by the simple scheme shown in Table 2. The inhibitor binds to the same site on the enzyme as the substrate; substrate analogues and often one of

**Table 2** Kinetic behaviour of simple reversible inhibition

| Inhibitor type | Basic mechanism | Kinetic equations |
|---|---|---|
| **Competitive** | $E + S \rightleftharpoons E.S \rightarrow E + P$; $E + I \rightleftharpoons E.I$ ($K_i$) | $v = \frac{V.\boldsymbol{s}}{\boldsymbol{s} + K_m(1 + \boldsymbol{i}/K_i)} = \frac{V}{1 + \frac{K_m}{\boldsymbol{s}}\left(1 + \frac{\boldsymbol{i}}{K_i}\right)}$ |
| **Uncompetitive** | $E + S \rightleftharpoons E.S \rightarrow E + P$; $E.S + I \rightleftharpoons E.S.I$ ($K_i$) | $v = \frac{\frac{V.\boldsymbol{s}}{(1 + \boldsymbol{i}/K_i)}}{\frac{K_m}{(1 + \boldsymbol{i}/K_i)} + \boldsymbol{s}} = \frac{V}{1 + \frac{\boldsymbol{i}}{K_i}\left(1 + \frac{K_m}{\boldsymbol{s}}\right)}$ |
| **Non-competitive**[a] | $E + S \rightleftharpoons E.S \rightarrow E + P$ ($K_s$); $E + I \rightleftharpoons E.I$ ($K_i$); $E.S + I \rightleftharpoons E.S.I$ ($K_i$); $E.I + S \rightleftharpoons E.S.I$ ($K_s$) | $v = \frac{\frac{V.\boldsymbol{s}}{(1 + \boldsymbol{i}/K_i)}}{K_s + \boldsymbol{s}} = \frac{V}{\left(1 + \frac{K_s}{\boldsymbol{s}}\right)\left(1 + \frac{\boldsymbol{i}}{K_i}\right)}$ |
| **Mixed**[a] | $E + S \rightleftharpoons E.S \rightarrow E + P$ ($K_s$); $E + I \rightleftharpoons E.I$ ($K_i$); $E.S + I \rightleftharpoons E.S.I$ ($K'_i$); $E.I + S \rightleftharpoons E.S.I$ ($K'_s$) | $v = \frac{V.\boldsymbol{s}/(1 + \boldsymbol{i}/K'_i)}{\boldsymbol{s} + K_s.\frac{(1 + \boldsymbol{i}/K_i)}{(1 + \boldsymbol{i}/K'_i)}} = \frac{V}{(1 + \boldsymbol{i}/K'_i) + \left(1 + \frac{K_s}{\boldsymbol{s}}\right)\left(1 + \frac{\boldsymbol{i}}{K_i}\right)}$ |

***s*** and ***i*** represent the concentrations of substrate and inhibitor, respectively. $K_s$, $K'_s$, $K_i$ and $K'_i$ are dissociation constants for the steps indicated. Since substrate and inhibitor binding steps in the noncompetitive and mixed cases are at equilibrium: $K_i.K'_s = K'_i.K_s$

[a] In the noncompetitive and mixed cases omission of the $E.I \rightleftharpoons E.S.I$ step gives the same equations.

**Table 3** Effects of inhibitor and substrate concentrations in reversible inhibition

| Inhibitor type | % Inhibition | $IC_{50}$ |
|---|---|---|
| **Competitive** | $\dfrac{100}{1 + \dfrac{K_i}{i}\left(1 + \dfrac{s}{K_m}\right)}$ | $K_i\left(1 + \dfrac{s}{K_m}\right)$ |
| **Uncompetitive** | $\dfrac{100}{1 + \dfrac{K_i}{i}\left(1 + \dfrac{K_m}{s}\right)}$ | $K_i\left(1 + \dfrac{K_m}{s}\right)$ |
| **Non-competitive** | $\dfrac{100}{1 + \dfrac{K_i}{i}}$ | $K_i$ |
| **Mixed** | $\dfrac{100}{\dfrac{\dfrac{(K_s + s)}{i} + \dfrac{K_s}{K_i} + \dfrac{s}{K'_i}}{\dfrac{K_s}{K_i} + \dfrac{s}{K'_i}}}$ | $\dfrac{s + K_s}{\dfrac{s}{K_i} + \dfrac{K_s}{K_i}}$ |

the products of enzyme-catalysed reactions are competitive inhibitors. The enzyme can either bind substrate **or** inhibitor; it cannot bind both at once. At very high substrate concentrations all inhibitor will be displaced from the enzyme, therefore the limiting velocity, $V$, will not be changed but more substrate will be needed to attain $V$ in the presence of the inhibitor. Thus more substrate will be needed to get to $V/2$. So for competitive inhibitors $V$ is unchanged but $K_m$ is increased, as shown mathematically in Table 2 and diagrammatically in Figures 2 and 5.

Pharmacologists frequently use $IC_{50}$ (or $I_{50}$), the inhibitor concentration required to give 50% inhibition, as a measure of potency. However, the measured $IC_{50}$ value depends on the substrate concentration, as shown in Table 3. At any fixed inhibitor concentration the degree of inhibition will decrease as the substrate concentration (**s**) is increased, tending to zero as **s** becomes very large, as shown in Figure 4.

Many reversible enzyme inhibitors, and most that are used as drugs, have been designed to look like a substrate (or product) for the enzyme and thus to be competitive inhibitors. Table 4 lists some common reversible inhibitors. Although the types of inhibition and $K_i$ values are shown, these may be affected by the assay conditions and substrates used, the species and tissue from which the enzyme was obtained, and whether a specific isoenzyme is being studied. Zollner (23) has published a relatively comprehensive compendium of enzyme inhibitors. Although competitive inhibitors are generally considered to bind to the same site on the enzyme as the substrate, it is also possible that the inhibitor might bind to a distinct site on the enzyme and cause a structural change that prevents substrate binding.

### 5.1.2 Competitive substrates

Some compounds that have been used as competitive inhibitors of enzymes are, in fact, alternative substrates for them. The situation where two substrates

**Table 4** Some reversible enzyme inhibitors

| Enzyme | Inhibitor | $K_i$ (μM) | Reference |
|---|---|---|---|
| **(a) Competitive** | | | |
| Acetylcholinesterase (EC 3.1.1.7) | 2-Hydroxybenzyltrimethylammonium | 0.02 | Wilson and Quan (1958) |
| Aconitate hydratase (EC 4.2.1.3) | Fluorocitrate | 200 | Peters and Wilson (1952) |
| Alanine dehydrogenase (EC 1.4.1) | D-Alanine | 20 000 | Yoshida and Freese (1970) |
| Aldehyde dehydrogenase (EC 1.12.1.3) | Chloralhydrate | 5 | Weiner *et al.* (1982) |
| Arginase (EC 3.5.3.1) | Adenine | 14 | Rosenfield *et al.* (1975) |
| Ascorbate peroxidase (EC 1.11.1.11) | Cyanide (towards $H_2O_2$) | 8300 | Shigeoka *et al.* (1980) |
| Carboxypeptidase A (EC 3.4.17.1) | Indole-3-acetic acid | 78 | Smith *et al.* (1951) |
| Chymotrypsin (EC 3.4.21.1) | Indole | 720 | Cunningham (1965) |
| Creatine kinase (EC 2.7.3.2) | Tripolyphosphate | 8 | Noda *et al.* (1960) |
| Dihydropteroate synthase (EC 2.5.1.15) | 4-Aminobenzoylglutamate | 1000 | Richey and Brown (1971) |
| Isocitrate dehydrogenase (NAD) (EC 1.1.1.41) | 3 Mercapto-2-oxoglutarate | 0.52 | Plaut *et al.* (1986) |
| Isocitrate dehydrogenase (NADP) (EC 1.1.1.42) | 3-Mercapto-2-oxoglutarate | 0.005 | Plaut *et al.* (1986) |
| Monoamine oxidase-A (EC 1.4.3.4) | D-Amphetamine | 20 | Mantle *et al.* (1976) |
| Pepsin A (EC 3.4.23.1) | Benzamide | 18 | Mares-Guia and Shaw (1965) |
| Succinate dehydrogenase (EC 1.3.99.1) | Malonate | 40 | Kearney (1957) |
| Tryptophanase (EC 4.1.99.1) | Indole | 50 | Newton *et al.* (1965) |
| Tyrosine aminotransferase (EC 2.6.1.5) | 3,5,3′ - Triiodothyronine | 38 | Diamondstone and Litwack (1963) |
| **(b) Uncompetitive** | | | |
| Alanopine dehydrogenase (EC 1.5.1.17) | 2-Oxoglutarate (towards NADH) | 1300 | Fields and Hochachka (1981) |
| Ascorbate peroxidase (EC 1.11.1.11) | Cyanide (towards ascorbate) | 8300 | Shigeoka *et al.* (1980) |
| Inositol-polyphosphate 1-phosphatase (EC 3.1.3.57) | Lithium ion | 6000 | Inhorn and Majerus (1987) |
| **(c) Mixed or non-competitive** | | | |
| *S*-Adenosylmethoionine decarboxylase (EC 4.1.1.50) | Spermine | 500 | Sakai *et al.* (1979) |
| Alanopine dehydrogenase (EC 1.5.1.17) | Succinate (towards NADH) | 1000 | Rubenstein *et al.* (1975) |
| Aldehyde reductase (EC 1.1.1.21) | Aldrestatin | 260 | Davidson *et al.* (1985) |
| α-Amylase (EC 3.2.1.1) | Acarbose | 1500 | De Mot and Verachtert (1987) |
| Arginase (EC 3.5.3.1) | Cytidine | 5 | Rosenfield *et al.* (1975) |
| *d*-CMP deaminase (EC 3.5.4.12) | TTP | 100 | Maley and Maley (1964) |
| Glycine-tRNA ligase (EC 6.1.1.14) | Serine | 52000 | Boyko and Fraser (1964) |
| Isocitrate lyase (EC 4.1.3.1) | Oxalate | 2000 | Smith and Gunsalus (1958) |
| Monophenol monooxygenase (EC 1.14.18.1) | l-Methyl-2-mercaptoimidazole | 4.6 | Andrawis and Khan (1986) |
| NAD synthase (EC 6.3.5.1) | Psicofuranine | 600 | Spencer and Priess (1967) |
| Phenylalanine-4-monooxygenase (EC 1.14.16.1) | Methotrexate | 38 | Craine *et al.* (1972) |
| Pyruvate dehydrogenase (lipoamide) (EC 1.2.4.1) | Isobutyryl-CoA | 90 | Martin-Requero *et al.* (1983) |
| | Methylmalonyl-CoA | 350 | Martin-Requero *et al.* (1983) |
| Serine acetyltransferase (EC 2.3.1.30) | L-Cysteine | 0.6 | Smith and Thompson (1971) |

Products of the reactions, whose inhibition type will depend on the kinetic mechanism followed (see Table 7), are not included.
$K_i$ values will depend on the assay conditions and perhaps also on the source of the enzyme.
In the case of the mixed inhibitors the $K_i$ values shown are those determined from the dependence of the apparent ($K_m/V$) values (slopes of double-reciprocal plots) upon the inhibitor concentration.

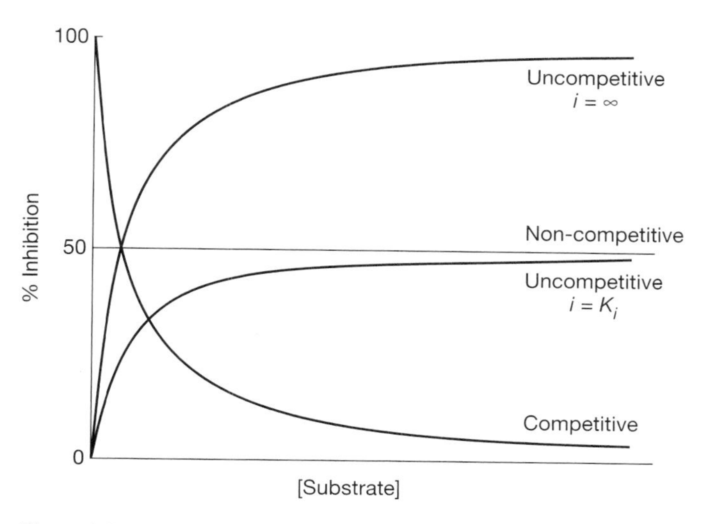


**Figure 4** The effects of substrate concentration on the degree of inhibition by reversible inhibitors. Unless otherwise indicated each inhibitor is at its $K_i$ concentration.


Andrawis, A. and Khan, V. (1986). *Biochem. J.*, **236**, 91.

Boyko, J. and Fraser, M. J. (1964). *Can J. Biochem.*, **42**, 1677.

Craine, J. E., Hall, E. S., and Kauffman, S. (1972). *J. Biol. Chem.*, **247**, 6082.

Cunningham, L. (1965). *Comp. Biochem.*, **16**, 85.

Davidson, W. S. and Murphy, D. G. (1985). *Progr. Clin. Biol. Res.*, **174**, 251.

De Mot, R. and Verachtert, H. (1987). *Eur. J. Biochem.*, **164**, 643.

Diamondstone, T. I. and Litwack, G. (1963). *J. Biol. Chem.*, **238**, 3859.

Fields, J. H. A. and Hochachka, P. W. (1981). *Eur. J. Biochem.*, **114**, 615.

Inhorn, R. C. and Majerus, P. W. (1987). *J. Biol. Chem.*, **262**, 15956.

Kearney, E. B. (1957). *J. Biol. Chem.*, **229**, 363.

Maley, G. F. and Maley, F. (1964). *J. Biol. Chem.*, **239**, 1168.

Mantle, T. J., Tipton, K. F., and Garrett, N. J. (1976). *Biochem. Pharmacol.*, **25**, 2073.

Mares-Guia, M. and Shaw, E. (1965). *J. Biol. Chem.*, **240**, 1579.

Martin-Requero, A., Corkey, B. E., and Cerdan, S. (1983). *J. Biol. Chem.*, **258**, 15338.

Newton, W. A., Morina, Y., and Snell, E. E. (1965). *J. Biol. Chem.*, **240**, 1211.

Noda, L., Nihl, T., and Morales, M. F. (1960). *J. Biol. Chem.*, **235**, 2830.

Peters, R. A. and Wilson, T. H. (1952). *Biochim. Biophys. Acta.*, **9**, 310.

Plaut, G. W. E., Aogaichi, T., and Gabriel, J. L. (1986). *Arch. Biochem. Biophys.*, **245**, 114.

Richey, D. P. and Brown, G. M. (1971). *Methods in enzymology*, **18B** (ed. D. B. McCormick), p. 771. Academic Press, New York.

Rosenfield, J. L., Dutta, S. P., Chednag, B., and Tritsch, G. L. (1975). *Biochim. Biophys. Acta*, **410**, 164.

Rubenstein, J. A., Collins, M. A., and Tabakoff, B. (1975). *Experientia*, **31**, 414.

Sakai, T., Hori, C., Kano, K., and Oka, T. (1979). *Biochemistry*, **18**, 5541.

Shigeoka, S., Nakano, Y., and Kitaoka, S. (1980). *Arch. Biochem. Biophys.*, **201**, 121.

Smith, E. L., Lumry, R., and Polglase, W. J. (1951). *J. Phys. Colloid Chem.*, **55**, 125.

Smith, I. K. and Thompson, J. F. (1971). *Biochim. Biophys. Acta*, **227**, 288.

Smith, R. A. and Gunsalus, I. C. (1958). *J. Biol. Chem.*, **229**, 305.

Spencer, R. L. and Priess, J. (1967). *J. Biol. Chem.*, **242**, 385.

Weiner, H., Freytag, S., Fox, J. M., and Hu, J. H. (1982). *Progr. Clin. Biol. Res.*, **114**, 91.

Wilson, I. B. and Quan, C. (1958). *Arch. Biochem. Biophys.*, **73**, 131.

Yoshida, A. and Freese, E. (1970). *Methods in enzymology*, **17** (ed. H. Tabor and C. W. Tabor), p. 176. Academic Press, New York.

compete for the same enzyme is also quite common in metabolic systems. Such competition can be represented by the simple mechanism in which substrates A and B are converted to products P and Q, respectively, by the same enzyme (E):

$$\begin{array}{ccc} & & E.A. \xrightarrow{k^a_{cat}} E + P \\ & k_1\boldsymbol{a} \nearrow \swarrow k_{-1} & \\ E & & \\ & k_{-3} \nwarrow \searrow k_3\boldsymbol{b} & \\ & & E.B. \xrightarrow{k^b_{cat}} E + Q \end{array} \qquad [9]$$

In this case, each substrate will act as a competitive inhibitor of the other. For example, if the initial rates of the formation of P ($v_a$) are measured at a series of different concentrations of A in the presence of different concentrations of B, the kinetic equation is of the same form as for simple competitive inhibition (see Table 2):

$$v_a = \frac{k^a_{cat}\,\mathbf{e}.\boldsymbol{a}}{\boldsymbol{a} + K^a_m\left(1 + \dfrac{\boldsymbol{b}}{K^b_m}\right)} = \frac{k^a_{cat}\,\mathbf{e}.\boldsymbol{a}/K^a_m}{1 + \dfrac{\boldsymbol{a}}{K^a_m} + \dfrac{\boldsymbol{b}}{K^b_m}} \qquad [10]$$

where $\boldsymbol{a}$ and $\boldsymbol{b}$ are the concentrations of A and B, respectively, $\boldsymbol{e}$ is enzyme concentration, and $K_m{}^a$ and $K_m{}^b$ are the Michaelis constants for A and B ($\{k_{-1} + k^a{}_{cat}\} / k_{+1}$ and $\{k_{-2} + k^b{}_{cat}\} / k_{+2}$, respectively). Similarly, the initial rates of Q formation ($v_b$) in the presence of A will be:

$$v_b = \frac{k^b_{cat}\,\mathbf{e}.\boldsymbol{b}}{\boldsymbol{b} + K^b_m\left(1 + \dfrac{\boldsymbol{a}}{K^a_m}\right)} = \frac{k^b_{cat}\,\mathbf{e}.\boldsymbol{b}/K^b_m}{1 + \dfrac{\boldsymbol{a}}{K^a_m} + \dfrac{\boldsymbol{b}}{K^b_m}} \qquad [11]$$

From Equations 10 and 11 the rates at which the enzyme will use each substrate in a mixture of the two will be given by the relative substrate concentrations and the ratio of their $k_{cat}/K_m$ (or, since $V = k_{cat}\mathbf{e}$), $V/K_m$ values:

$$\frac{v_a}{v_b} = \frac{(k^a_{cat}/K^a_m)\boldsymbol{a}}{(k^b_{cat}/K^b_m)\boldsymbol{b}} = \frac{(V_a/K^a_m)\boldsymbol{a}}{(V_b/K^b_m)\boldsymbol{b}} \qquad [12]$$

The relationships also provide a way of determining whether the same enzyme is involved in the metabolism of two different substrates. If each substrate is assayed at its $K_m$ concentration ($\boldsymbol{a} = K_m{}^a$ and $\boldsymbol{b} = K_m{}^b$ ) $\boldsymbol{a}/K_m{}^a$ and $\boldsymbol{b}/K_m{}^b$ will be unity. The rates for each substrate assayed separately will each be one-half their limiting velocities and the sum of the two rates would be:

$$v_a + v_b = (V_a/2) + (V_b/2) = \frac{V_a + V_b}{2} \qquad [13]$$

If rate of formation of both products ($v_{(a\,+\,b)}$) from mixture of the two substrates, each at its $K_m$ concentration, is determined, the above relationships will be true if they were metabolized by separate enzymes. However, if the same enzyme were responsible for the metabolism of both the substrates, substitution into Equations 10 and 11 shows that the rate would be:

$$v_{(a+b)} = (V_a/3) + V_b/3) = \frac{V_a + V_b}{3} \qquad [14]$$

and so the rate with the mixture will be two-thirds of the sum of the separate rates under the same conditions.

Some competitive substrates of physiological significance are shown in Table 5.

### 5.1.3 Uncompetitive inhibition

This is less common than competitive inhibition but it does occur, particularly in the case of enzyme reactions involving more than one substrate. The simplest model is that shown in Table 2. The kinetic equation in Table 2 shows that both $K_m$ and $V$ are decreased by the same factor, resulting in the behaviour shown in Figures 2 and 5. Competitive and uncompetitive inhibition are easily distinguished from plots of their kinetic behaviour. The characteristic double-reciprocal plots resulting from these two forms of inhibition are shown in Figure 5. The relationship between the $IC_{50}$ and the substrate concentration is shown in Table 3. Inhibition will increase as the concentration of substrate is increased with the $IC_{50}$ value tending to $K_i$ as substrate concentration tends to infinity. This is the exact opposite of the competitive case (Figure 4).

### 5.1.4 Mixed and non-competitive inhibition

In this case the inhibitor can bind both to free enzyme and to the enzyme–substrate complex. Since this system involves alternative pathways (see Section 4.2), the steady-state kinetic analysis leads to a complex equation that predicts non-hyperbolic dependence of initial velocity upon substrate or inhibitor concentration. However, if it is assumed that the rate of breakdown of the E.S complex to give products is much slower than the rates of dissociation of the complexes, all the binding steps can be regarded as equilibria. Under such conditions $K_s$, $K'_s$, $K_i$, and $K'_i$ will all be simple dissociation constants and the kinetic equation describing the system can be written as shown in Table 2. Because the system is in equilibrium only three of the dissociation constants are necessary to define the system ($K_i/K'_i = K_s/K'_s$).

The kinetic behaviour looks like a mixture of competitive and uncompetitive inhibition. Thus the double-reciprocal plot can give any of the patterns shown in Figure 5. The case where $K'_i = K_i$ (and therefore $K_s = K'_s$) is often known as non-competitive inhibition but confusion can arise since some authors use this term to cover mixed inhibition as well.

The dependence of $IC_{50}$ on substrate concentration will depend on the relative values of $K_i$ and $K_i'$: it can either increase, be unaffected (non-competitive inhibition), or decrease with **s** (see Table 3 and Figure 4).

### 5.1.5 Determination of inhibitor constants

Table 6 summarizes the relationship between the apparent values of $V$, $K_m/V$, and $K_m$ ($V^{app}$, $K_m{}^{app}/V^{app}$, $K_m{}^{app}$, respectively) for the different types of inhibition. In the

**Table 5** Some competing substrates

| Enzyme | Substrate | Competitor | Comments |
|---|---|---|---|
| Alcohol dehydrogenase (EC 1.1.1.1) | Ethanol | Allyl alcohol | Ingested allyl alcohol is metabolized to hepatotoxic acrolein; ethanol consumption may protect by competition[a] |
| | Retinol | Ethanol | See below[b] |
| Aldehyde dehydrogenase (EC 1.12.1.3) | 'Biogenic aldehydes' | Acetaldehyde | Acetaldehyde derived from ethanol decreases oxidation of the aldehyde derived from noradrenaline to its corresponding acid[c] |
| | Retinal | Acetaldehyde | Some of the effects of alcoholism may involve interference with the metabolism of retinol and its derivatives[b] |
| Dopa decarboxylase (EC 4.1.1.28) | Dopa | α-Methyldopa[d] | This compound was introduced as an enzyme inhibitor but some of its effects result from the formation of the α-methyl analogues of dopamine and noradrenaline which interfere with their functions[e] |
| Isoleucine-tRNA synthase (EC 6.1.1.5) | Isoleucine | Valine | Formation of incorrect aminoacyl-tRNAs by reaction of these poor competitors is prevented by 'editing' and 'proof-reading' mechanisms[f] |
| Valyl-tRNA synthase (EC 6.1.1.9) | Valine | Threonine | |

[a] Stoner, H. B. and McGee, P. N. (1957). *Br. Med. Bull.*, **13**, 102.
[b] Leo, M. A. and Lieber, C. S. (1983). *Alc. Clin. Exp. Res.*, **7**, 15.
[c] Turner, A. J., Illingworth, J. A., and Tipton, K. F. (1974). *Biochem. J.*, **114**, 353.
[d] Also gives time-dependent inhibition (see ref. 67).
[e] Gilman, A. G. G., Goodman, L. S., Rall, T. W., and Murad, F. (1985). *Goodman and Gillman's The pharmacological basis of therapeutics.* Macmillan, USA.
[f] Fersht, A. R. (1999). *Structure and mechanism in protein science: a guide to enzyme catalysis and protein folding.* W. H. Freeman & Co, USA.

past the slopes and intercepts from double-reciprocal plots (see Figure 5) have often been used to determine the apparent values of $K_m/V$ and $1/V$ in the presence of inhibitor and similarly the extrapolated baseline intercept has been used for determining the apparent value of $-1/K_m$. The $K_i$ values can then be obtained from plots of these values against the inhibitor concentration, as shown in Figure 5. However, as discussed above, direct fitting of the initial rate data and derived parameters by non-linear regression should be the method of choice since it is a considerably more accurate procedure. Although the double-reciprocal plot is known to be the least accurate of all methods for analysing inhibitor effects, it does remain a convenient, and widely-understood procedure for their display and is used for such purposes here.

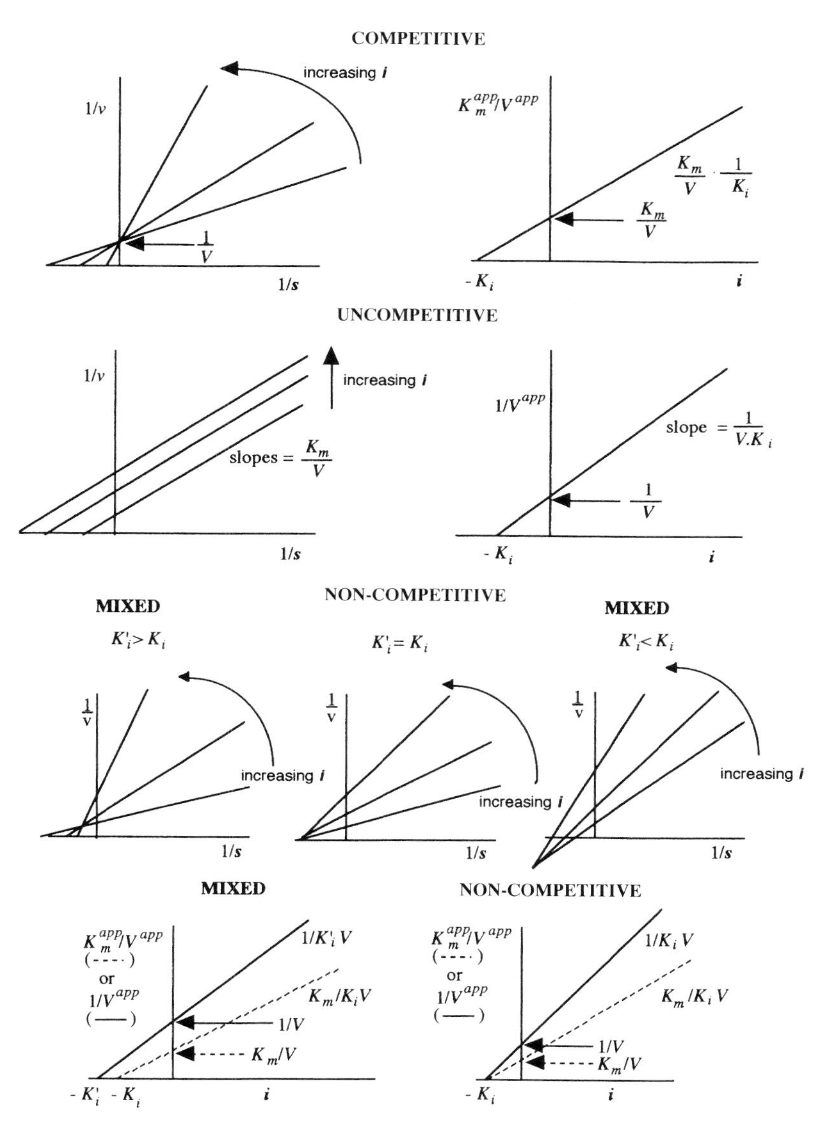


**Figure 5** Characteristic double-reciprocal plots obtained in the presence of the different types of reversible inhibitor.

An alternative graphical procedure for determining $K_i$ values, originally proposed by Dixon in 1953 (24, see also ref. 15), is still sometimes used. This involves plotting reciprocal velocity against inhibitor concentration ($1/v$ versus $\boldsymbol{i}$) at a series of fixed substrate concentrations. For competitive inhibition this will give a family of straight lines for plots of $1/v$ versus $\boldsymbol{i}$ at a series of substrate concentrations. Since inhibition is competitive these lines must intersect at a common point above the $-\boldsymbol{i}$ axis which will correspond to a value of $1/V$ on the

**Table 6** The apparent values of the Michaelis–Menten parameters in the presence of simple reversible inhibitors

| Inhibitor type | $K_m^{app}$ | $V^{app}$ | $(K_m^{app}/V^{app})$ |
|---|---|---|---|
| Competitive | $K_m(1 + \mathbf{i}/K_i)$ | $V$ | $K_m(1 + \mathbf{i}/K_i) / V$ |
| Uncompetitive | $K_m / (1 + \mathbf{i}/K_i)$ | $V / (1 + \mathbf{i}/K_i)$ | $K_m / V$ |
| Non-competitive | $K_m$ | $V / (1 + \mathbf{i}/K_i)$ | $K_m(1 + \mathbf{i}/K_i) / V$ |
| Mixed | $K_m(1 + \mathbf{i}/K_i) / (1 + \mathbf{i}/K_i')$ | $V / (1 + \mathbf{i}/K_i')$ | $K_m(1 + \mathbf{i}/K_i) / V$ |

$1/v$ axis and value of this common intersection point above the above the $-\boldsymbol{i}$ axis will be $-K_i$, as shown in Figure 6.

Provided it is established that the inhibitor is competitive and the maximum velocity is accurately known, it is possible, although not very accurate, to determine $K_i$ from the intersection between one experimental line and a line drawn parallel to the inhibitor concentration axis such that its height above that axis corresponds to $1/V$. It follows from the above arguments that the lines in a Dixon plot for a competitive inhibitor must always intersect above the $-\boldsymbol{i}$ axis. This has often been used as a diagnostic test for competitive inhibition but this is not valid since mixed inhibitors may also give families of lines that intersect in the same quadrant. This procedure has been extended to the analysis of mixed and uncompetitive inhibition but since it offers no advantages over direct, non-linear regression, fitting of the data it will not be considered further.

### 5.1.6 Product inhibition

As discussed above, the products of enzyme-catalysed reactions are frequently inhibitors of the reaction. For a reaction involving the transformation of a single substrate to product (see Table 2) the product P would be expected to be a simple competitive inhibitor of the initial rate of the reaction in the forward direction. Most enzyme-catalysed reactions involve the transformation of two, or more, substrates into two, or more, products and the types of product inhibition observed can be useful in determining the order of substrate binding to, and

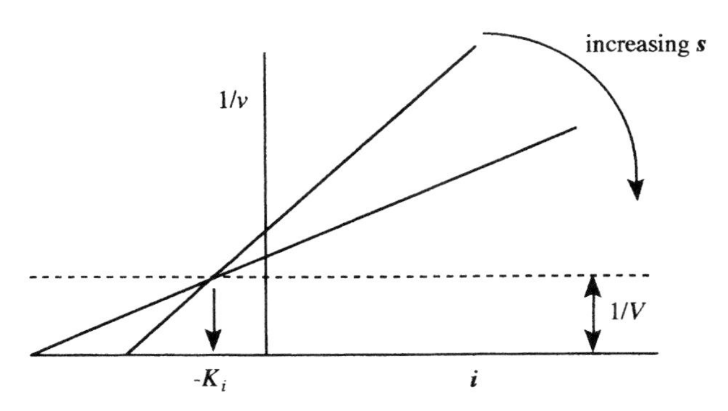


**Figure 6** The Dixon plot for determining the inhibitor constant ($K_i$) for a competitive inhibitor.

product release from, the enzyme. Table 7 shows the types of product inhibition that may be encountered for simple two-substrate/two-product reactions, to illustrate that each of the different types of inhibition discussed above may be found. Fuller accounts of the kinetics of enzymes involving two substrates can be found in refs 14–16 and 25–28. Detailed discussions of the kinetic behaviour and product inhibition of reaction mechanisms involving three (16, 28–30) and four (31) substrates have also been presented. In systems with more than one substrate analysis of the kinetic behaviour of competing substrates (see Section 5.1.2) can also be used to reveal details of the reaction pathway involved (see ref. 32).

**Table 7** Simple product inhibition patterns for two-substrate / two-product reactions of the form:

$$Ax + B \rightleftharpoons A + Bx$$

| **Mechanism** | **Product** | **Inhibition towards Ax** | | **Inhibition towards B** | |
|---|---|---|---|---|---|
| | | ***B*** not saturating | ***B*** saturating | ***Ax*** not saturating | ***Ax*** saturating |
| Ordered | A | Competitive | Competitive | Mixed | None |
| | Bx | Mixed | Uncompetitive | Mixed | Mixed |
| Theorell–Chance | A | Competitive | Competitive | Mixed | None |
| | Bx | Mixed | None | Competitive | Competitive |
| Double-displacement | A | Mixed | None | Competitive | Competitive |
| | Bx | Competitive | Competitive | Mixed | None |
| Random-order | A | Competitive | Competitive | Mixed | None |
| | Bx | Mixed | None | Competitive | Competitive |

In the ordered mechanism the enzyme binds substrate Ax to form an E·Ax binary complex that subsequently binds B to give the E·Ax·B succeeded by the E·A·Bx ternary complexes, product Bx is then released before dissociation of A.

The order of substrate binding and product release is the same for the Theorell–Chance mechanism, but in this case the concentrations of the E·Ax·B and E·A·Bx ternary complexes are not kinetically significant.

In the double-displacement (ping-pong) mechanism the E·Ax complex reacts and releases product A to give a modified form of the enzyme Ex, which subsequently binds B and reacts with it to regenerate E and release Bx.

The random-order mechanism involves the enzyme being able to bind its substrates in any order, Ax before B or B before Ax, to give the ternary E·Ax·B complex.

### 5.1.7 More complex inhibition behaviour

Apart from artefacts due to incorrect assay procedures, which can lead to apparently complex behaviour (see ref. 6), a number of inhibitor interactions can show departure from simple Michaelis–Menten kinetics.

#### *i. Partial reversible inhibition*

It is possible that an inhibitor may not completely prevent the enzyme-catalysed reaction from occurring. For example in the competitive case (see Table 2) a situation might exist where the binding of inhibitor did not prevent substrate from binding to the enzyme, but reduced the affinity of such binding without affecting the rate of product formation from the E.S or E.I.S complex, as shown below.

$$
\begin{array}{ccccc}
 & K_s & & k_{cat} & \\
E + S & \rightleftharpoons & E.S. & \longrightarrow & E + \text{Products} \\
K_i \updownarrow + I & & Ki' \updownarrow + I & & \\
 & & & k'_{cat} & \\
E.I & \rightleftharpoons & E.S.I. & \longrightarrow & E.I. + \text{Products} \\
 & K'_s & & &
\end{array} \quad [15]
$$

Partially competitive inhibition will occur if $k_{cat} = k'_{cat}$ but $K'_s > K_s$ and therefore $K'_1 > K_i$. (Note if $k_{cat} = k'_{cat}$ but $K'_s < K_s$ and therefore $K'_1 < K_i$ partial competitive activation will occur.) As in the case of mixed inhibition steady-state treatment of such a system will yield a complex equation (see refs 15 and 33). If one assumes equilibrium conditions to apply for all the binding steps the initial rate equation, the resulting kinetic equation is:

$$
v = \frac{V.\mathbf{s}}{\mathbf{s} + K_s\left(\dfrac{1 + \mathbf{i}/K_i}{1 + \dfrac{\mathbf{i}}{K_i}\dfrac{K_s}{K'_s}}\right)} = \frac{V.\mathbf{s}}{\mathbf{s} + K_s\left(\dfrac{1 + \mathbf{i}/K_i}{1 + \mathbf{i}/K'_i}\right)} \quad [16]
$$

When $\mathbf{i} = 0$ this relationship reduces to the simple Michaelis–Menten equation $v = V\mathbf{s} / (\mathbf{s} + K_s)$ and if $\mathbf{i}$ is very large the relationship becomes $v = V\mathbf{s} / (\mathbf{s} + K'_s)$. In this case the inhibitor will not completely prevent the reaction from occurring, since at very high inhibitor concentrations reaction will still occur by substrate binding to the E.I complex. Since the rates of breakdown of the E.S and E.S.I complexes to yield products are, by definition the same, it also follows that the maximum velocity that will be obtained at sufficiently high substrate concentrations will be unchanged by the presence of the inhibitor. However a graph of the apparent $K_s/V$ ($K_s^{app}/V^{app}$) against $\mathbf{i}$ will curve downward towards a limiting value of $K'_s/V$. Dixon plots of $1/\mathbf{v}$ against $\mathbf{i}$ will also curve downwards.

The $K_s^{app}/V^{app}$ against $\mathbf{i}$ curve is a hyperbola that does not pass through the origin (Figure 7). Thus, if the baseline is shifted up to the point where this curve cuts the vertical axis the data can be fitted to the equation for a rectangular hyperbola in the normal way. If we define the change in slope value from the axis intersection point as $\Delta(K_s^{app}/V^{app})$, the relationships becomes:

$$
\Delta\left(\frac{K_s^{app}}{V^{app}}\right) = \frac{K_s}{V}\left(\frac{1 + \mathbf{i}/K_i}{1 + \mathbf{i}/K'_i}\right) - \frac{K_s}{V} = \frac{K_s\mathbf{i}/K_i - K_s\mathbf{i}/K'_i}{V + V\mathbf{i}/K'_i} \quad [17]
$$

This can, perhaps, be seen more clearly in the double-reciprocal form transformation:

$$
\Delta\left(\frac{V^{app}}{K_s^{app}}\right) = \frac{VK'_iK_i}{K_s(K'_i - K_i)} \cdot \frac{1}{\mathbf{i}} + \frac{VK_i}{K_s(K'_i - K_i)} \quad [18]
$$

which shows that a graph of $1/\Delta(K_s^{app}/V^{app})$ against $1/\mathbf{i}$ will be linear and intersect the horizontal axis at a value corresponding to $-1/K_i$ (see Figure 7).

Equation 15 will also describe partial mixed inhibition with the additional inequality of $k_{cat} > k'_{cat}$ and partial non-competitive inhibition if $k'_{cat} = 0$. In the

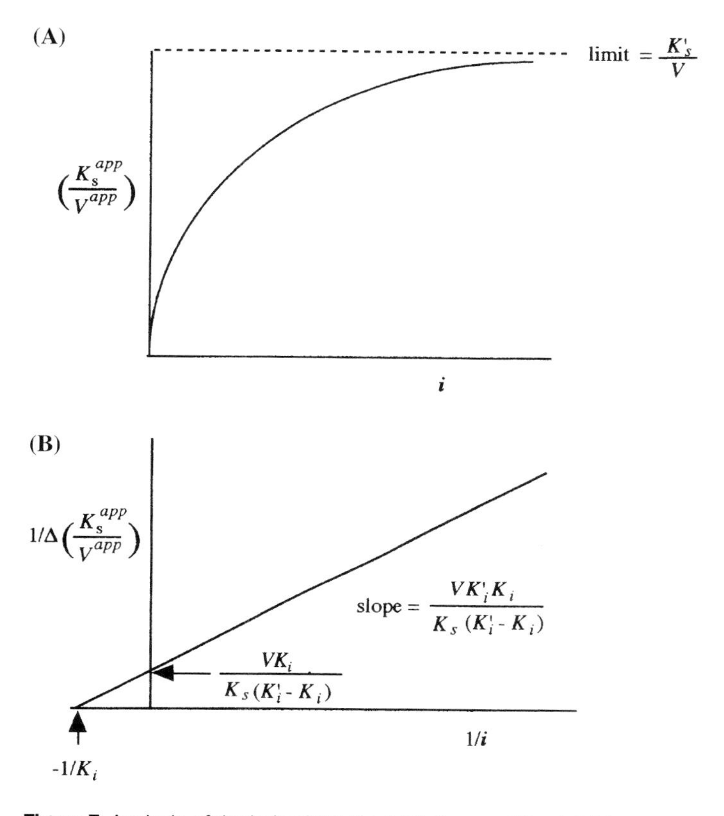


**Figure 7** Analysis of the behaviour of a partial competitive inhibitor.

special case where $K_i/K'_i = k_{cat}/k'_{cat}$, that is the ratio of the dissociation constants for the release of S from E.S and E.I.S is equal to the ratio of the rates of reaction of these two complexes to yield products, partial uncompetitive inhibition results. The kinetic behaviour in these cases can be analysed by similar procedures to the partial competitive case.

It might also be expected that partial uncompetitive inhibition would result if the inhibitor did not bind to the free enzyme and $k_{cat} > k'_{cat}$, as shown in Equation 19.

$$\begin{array}{l} E + S \overset{K_s}{\rightleftharpoons} E.S. \overset{K_{cat}}{\longrightarrow} E + Products \\ \qquad\qquad Ki' \updownarrow + I \\ \qquad\qquad E.S.I. \overset{k'_{cat}}{\longrightarrow} E.I. + Products \end{array} \qquad [19]$$

However, the kinetic behaviour does not resemble uncompetitive inhibition at all. In fact it is unclear whether it should be regarded as inhibition or activation, since $\boldsymbol{i}$ activates at low substrate concentrations and inhibits at higher ones

(Figure 8). At very high concentrations of inhibitor ($\boldsymbol{i} \rightarrow \infty$) the equilibrium between enzyme and E.S complex will all be displaced into E.S.I and therefore the apparent $K_s \rightarrow 0$, and $V$ will become constant at $k'_{cat}\boldsymbol{e}$. The kinetic equation, shown in Equation 20, where $V$ and $V'$ are $k_{cao}\boldsymbol{e}$ and $k'_{cao}\boldsymbol{e}$, respectively, shows that the value of $K_s^{app}/V^{app}$ against $\boldsymbol{i}$ will decrease to zero and the value of $1/V^{app}$ will decrease hyperbolically towards a value of $1/V'$ as $\boldsymbol{i}$ is increased.

$$v = \frac{V}{1 + \frac{K_s}{\boldsymbol{s}} + \frac{\boldsymbol{i}}{K_i}} + \frac{V'}{1 + \frac{K_i}{\boldsymbol{i}}\left(1 + \frac{K_s}{\boldsymbol{s}}\right)} = \frac{V}{\left(\frac{K_s}{1 + \frac{(V'/V)\boldsymbol{i}}{K_i}}\right)\frac{1}{\boldsymbol{s}} + \left(\frac{(1 + \boldsymbol{i}/K_i)}{1 + \frac{(V'/V)\boldsymbol{i}}{K_i}}\right)} \quad [20]$$

### *ii. Steady-state systems with alternative pathway*

As discussed in Section 4.2 steady-state kinetic treatment of systems in which there are alternative pathways leading to enzyme–substrate complexes that subsequently react to give product(s), such as the two-substrate reaction of Equation 5, mixed and non-competitive inhibition (Table 2), and the partial competitive and mixed inhibition cases shown in Equation 15 yields complex initial rate equations, of the form shown in Equations 6 and 7. Thus the addition of a fixed quantity of an inhibitor obeying such a mechanism can induce complex kinetics with respect to substrate conversion. Only in the absence of inhibitor will the behaviour reduce to that described by the simple Michaelis–Menten equation.

### *iii. Cooperativity*

The cooperative binding of many allosteric inhibitors will lead to kinetic behaviour that does not follow the simple Michaelis–Menten predictions as discussed in Section 4.2 (see also refs 11, 12, 14, and 25).

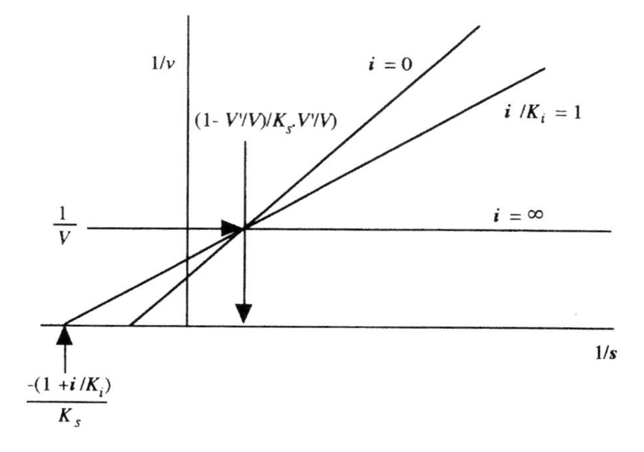


**Figure 8** The behaviour of a 'partial uncompetitive inhibitor' according to Equations 18 and 19.

### *iv. More than one inhibitor-binding site*

Systems in which more than one molecule of inhibitor can bind to the same form of the enzyme will lead to kinetic equations in which the inhibitor concentration is raised to a power (see refs 15 and 25). The equations given, under both equilibrium and steady-state conditions are similar, for example the simple extension of competitive inhibition:

$$\begin{array}{ccccc} \mathrm{E} & \underset{k_{-1}}{\overset{K_{+1}.\boldsymbol{s}}{\rightleftharpoons}} & \mathrm{E.S.} & \xrightarrow{k_{cat}} & \mathrm{E + Products} \\ K_i \updownarrow + \mathrm{I} & & & & \\ \mathrm{E.I} & \overset{+\mathrm{I}}{\rightleftharpoons} & \mathrm{E.I_2} & & \\ & & & K' & \end{array} \qquad [21]$$

will give an initial rate equation of the form:

$$v = \frac{V.\boldsymbol{s}}{\mathrm{s} + K_m\left(1 + \frac{\boldsymbol{i}}{K_i} + \frac{\boldsymbol{i}^2}{K'_i.K_i}\right)} \qquad [22]$$

which will yield normal kinetic behaviour with respect to substrate concentration, but curves of $K_m^{app}/V^{app}$ against $\boldsymbol{i}$ will be upwardly curving (parabolic). Clearly it is possible to devise more elaborate mechanisms than that shown. However, if the inhibitor binds to different enzyme intermediates the kinetic behaviour will remain simple. For example in the system:

$$\begin{array}{ccccc} \mathrm{E} & \underset{k_{-1}}{\overset{K_{+1}.\boldsymbol{s}}{\rightleftharpoons}} & \mathrm{E.S.} & \xrightarrow{k_{cat}} & \mathrm{E + Products} \\ K_i \updownarrow + \mathrm{I} & & K'_i \updownarrow + \mathrm{I} & & \\ \mathrm{E.I} & & \mathrm{E.S.I.} & & \end{array} \qquad [23]$$

the inhibitor will behave as a simple mixture of competitive and uncompetitive effects and the inhibition will appear to be mixed.

$$v = \frac{V.\boldsymbol{s}}{1 + K_m\left(1 + \frac{\boldsymbol{i}}{K_i}\right) + \frac{\boldsymbol{i}}{K'_i}} \qquad [24]$$

### *v. Inhibitor binds to substrate rather than enzyme*

Such a system might is represented in Equation 25, where L is the ligand that binds substrate and $K_d$ is the dissociation constant of the L.S complex..

$$\begin{array}{ccccc} \mathrm{E + S} & \underset{k_{-1}}{\overset{K_{+1}.}{\rightleftharpoons}} & \mathrm{E.S.} & \xrightarrow{k_{cat}} & \mathrm{E + Products} \\ K_d \updownarrow + \mathrm{L} & & & & \\ \mathrm{S.L} & & & & \end{array} \qquad [25]$$

In this system it is free, uncomplexed, substrate ($\mathbf{s}_f$ ) that can bind to the enzyme; the inhibitor simply reduces the amount of $\mathbf{s}_f$ available. The problem becomes that of knowing the free substrate concentration from the amounts of substrate and inhibitor added. As shown in Equation 26 the solution to this yields an unpleasant expression for the free substrate concentration which shows that in the presence of a fixed concentration of inhibitor the free substrate concentration will vary sigmoidally as the total substrate concentration is increased. In other words the substrate is being titrated against the inhibitor. Therefore, although the variation of velocity with free substrate concentration would obey normal Michaelis–Menten kinetics, the variation will describe a sigmoid curve with respect to total substrate concentration. Similarly, plots of velocity against total inhibitor concentration will also be sigmoid and plots of $1/v$ as a function of ligand concentration ($\mathbf{l}$) will be parabolic.

$$\mathrm{sf} = \frac{\sqrt{(\mathbf{l} - \mathbf{s} + K_d)^2 + 4\,K_d.\mathbf{s}} - (\mathbf{l} - \mathbf{s} + K_d)}{2} \qquad [26]$$

A simplifying situation will occur if the ligand concentration is so high that it is not significantly depleted by substrate binding ($\mathbf{l_f} \approx \mathbf{l}$) when the behaviour will be indistinguishable from competitive inhibition.

An alternative possibility is that the S.L complex is the true inhibitor. This situation can become very complicated since the velocity can be affected both by inhibition by S.L and by depletion of free substrate. Similar complexities will arise when activator forms a dissociable complex with the substrate (see ref. 10, and Section 5.1.8), as is often the case with metal cation activated enzymes.

### 5.1.8 Inhibition by high substrate concentrations

It is not uncommon for enzymes to be inhibited by high concentrations of one, or more, of their substrates, leading to kinetic plots such as those shown in Figure 9. Such behaviour can be useful in helping to deduce the kinetic mechanism involved (see refs 15, 16, 25, and 34) but it can restrict the range of substrate concentrations that can be used for determining $K_m$ and $V$ values.

In the simplest case it can be represented by the scheme:

$$\begin{array}{ccc} \mathrm{E} \underset{k_{-1}}{\overset{k_{+1}.\mathbf{s}}{\rightleftharpoons}} & \mathrm{E.S} & \xrightarrow{k_{cat}} \mathrm{E} + \mathrm{Products} \\ & \Big\updownarrow +\mathrm{S} & \\ & \mathrm{E.S_2} & \end{array} \qquad [27]$$

where binding of the second substrate molecule, to form $E.S_2$, prevents catalysis. This gives an equation of the form:

$$v = \frac{V\mathbf{s}}{K_m + \mathbf{s} + \dfrac{\mathbf{s}^2}{K_i^s}} = \frac{V}{1 + \dfrac{K_m}{\mathbf{s}} + \dfrac{\mathbf{s}}{K_i^s}} \qquad [28]$$

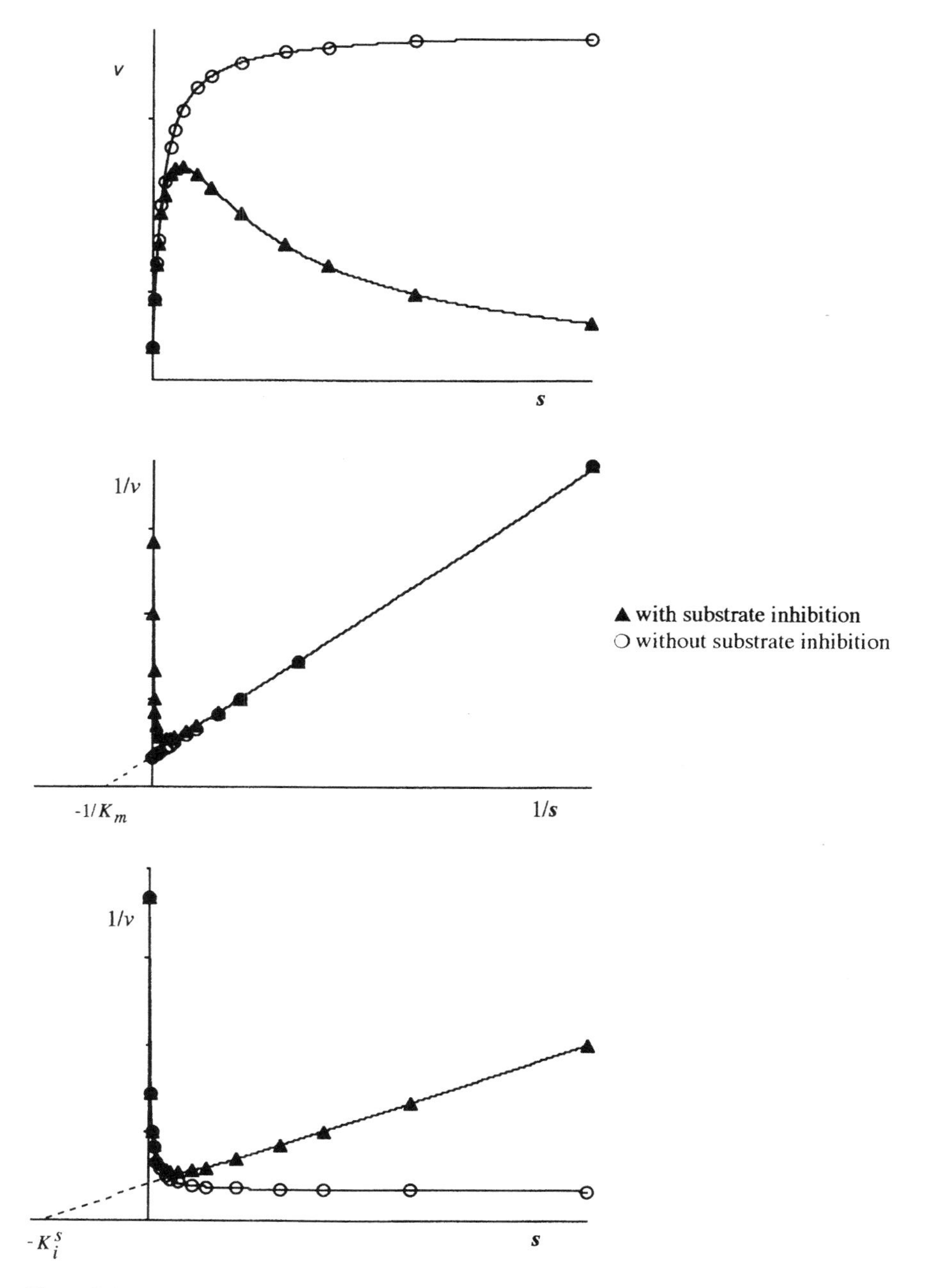


**Figure 9** Inhibition by high substrate concentrations.

Thus at lower substrate concentrations the variation of $v$ with **s** may appear to follow a simple rectangular hyperbola, the velocity will reach an apparent maximum and then decline as **s** is increased further. If $K_m$ and $K^S_I$ are sufficiently different (by a factor of 10 or more) the values of the two parameters may be obtained by directly fitting the initial rate behaviour to Equation 27. The kinetic behaviour is illustrated in Figure 9.

A special case of high-substrate inhibition can occur if the true substrate for the enzyme is a substrate–activator complex (see Section 5.1.7). This is frequently the case with metal ion-dependent enzymes; for example the magnesium–ATP complex is the true substrate for many kinases and the uncomplexed substrate may bind to the enzyme as an inhibitor. This will give rise to apparent high-substrate inhibition unless experiments are designed to assure that the metal ion concentrations are high enough to ensure that essentially all the substrate is in the complexed form over the entire concentration range used. Alternatively, there are some enzyme reactions that require metal ions binding to the enzyme as essential activators. In such cases reduction of the free metal ion concentration through chelation by one of the substrates can also result in inhibition. Cases where a complex between the enzyme and a metal ion is the active species and the true substrate is a complex between substrate and the same metal ion also occur, an example being pyruvate carboxylase (EC 6.4.1.1) (35). In such cases it is necessary to calculate the concentrations of metal–substrate complex and free metal ions, from the stability constants of the metal–ligand complexes in order to interpret the kinetic behaviour (see refs 10 and 36). The solution of the quadratic equations for cases where there are several metal-binding species would be a tedious procedure and computer solutions are available to do this (see e.g. ref. 37).

## 5.2 Tight-binding inhibitors

Some inhibitors have such high affinities for the enzyme that the concentrations required for inhibition are comparable to those of the enzyme. In such cases, the binding of the inhibitor to the enzyme will significantly reduce the free inhibitor concentration and so the assumption that the total inhibitor concentration is equal to the free inhibitor concentration, which has been implicit in all the above treatments, is not valid and it is necessary to derive relationships in terms of the concentrations of *free* enzyme and free inhibitor concentrations. For the reaction:

$$\mathrm{E} + 1 \rightleftharpoons \mathrm{E.I}$$

the dissociation constant ($K_s$) will be:

$$K_s = \frac{(\boldsymbol{e} - \boldsymbol{ei})(\boldsymbol{i} - \boldsymbol{ei})}{\boldsymbol{ei}} \qquad [29]$$

This relationship leads to an equation of the same form as Equation 26 and does not predict simple inhibition curves unless it is expressed in terms of free enzyme and inhibitor. Methods for analysing the behaviour of such systems have been presented elsewhere (38–40). The low concentrations of inhibitor (as well as those of the enzyme) used for inhibition studies may result in the rate of inhibition, governed by the second-order rate constant for combination between E and I, being relatively slow (see ref. 38).

Some tight-binding inhibitors are shown in Table 8. They are often analogues of the reaction transition state (see ref. 41). This can be represented as follows. A simple reaction:

$$Ax + B \rightleftharpoons A + Bx$$

must pass through a transition state in order to proceed:

$$Ax + B \rightleftharpoons \underset{\text{Transition state}}{A---X---B} \rightleftharpoons A + Bx$$

This can be written even more simply by representing the transition state is $(AB)^*$:

$$A + B \rightleftharpoons (AB)^* \rightleftharpoons P + Q$$

**Table 8** Some tight-binding enzyme inhibitors

| Enzyme | Substrate | $K_m$ (μM) | Analogue | $K_i$ (nM) |
|---|---|---|---|---|
| Carnitine *O*-acetyltransferase (EC 2.3.1.7) | AcetylCoA<br>$(CH_3)_3N^+CH_2CH(OH)CH_2OH$ (carnitine) | 34<br>120 | $(CH_3)_3N^+CH(CH_2COO^-)OC(=O)CH_2SCoA$ | <12 |
| Aspartate transcarbamoylase (EC 2.3.1.2) | Carbamoyl phosophate<br>Aspartate | 27<br>11 000 | $NHCOCH_2PO_3^{2-}$–$CHCOO^-$–$CH_2COO^-$ | 27 |
| Adenylate kinase (EC 2.7.4.3) | MgATP<br>AMP | 100<br>625 | Ado—OPOPOPOPOP—Ado.$Mg^{2+}$ (each P bearing $O^-$ and =O) | 2.5 |
| L-Lactate dehydrogenase (EC 1.1.1.27) | NADH<br>Pyruvate | 24<br>140 | Ado—OPOPOR—N (dihydropyridine ring bearing $CH_2COCOO^-$, H and $CONH^2$); each P bearing $O^-$ and =O | <1 |
| Cytidine deaminase (EC 2.1.6) | Cytidine (ring structure: NH, H–N, O, N–R) | 170 | Ring structure: HO, H, HO, O, N–R | 29 |
| Catechol *O*-methyltransferase (EC 3.5.4.5) | Catechol (benzene ring with two OH) | 32 | Benzene ring with OH, OH, $O_2N$, $NO_2$ | 8 |

Ado = adenosine; R = ribose.

The overall equilibrium of this process will be governed by the free energy change ($\Delta G_0$) since the equilibrium constant $K_{eq}$ for the overall reaction will be given by:

$$\Delta G_o = RT \ln \frac{1}{K_{eq}}$$

If the formation of the transition state is considered as an equilibrium:

$$K_{eq}{}^{*} = \frac{[A][B]}{[(AB)^{*}]} \therefore [(AB)^{*}] = \frac{[A][B]}{K_{eq}{}^{*}} \text{ and } \Delta G^{*} = \frac{RT}{Nh} \ln \frac{1}{K_{eq}{}^{*}} \quad [30]$$

and $\Delta G^{*}$ will be a measure of the difficulty in making the reaction occur.

Eyring showed that the overall reaction velocity will be given by:

$$v = \frac{RT}{Nh}[(AB^{*})] \text{ and } \therefore v = \frac{RT}{Nh}\frac{[A][B]}{K_{eq}{}^{*}} \quad [31]$$

The apparent overall second-order rate constant for the reaction leading to the transition state will be given by:

$$v = k[A][B] \quad [32]$$

and combining Equations 31 and 32:

$$k = \frac{RT}{Nh}\frac{1}{K_{eq}{}^{*}} \quad [33]$$

Thus fast reactions will have small values of $K_{eq}{}^{*}$. That is, they will have high concentrations of the transition state.

An enzyme is envisaged as accelerating a reaction by lowering the activation energy. It can do this by stabilizing the transition state using some of the energy derived from substrate binding to facilitate transition state formation. This is a variant on the 'substrate strain' theory. It predicts that an enzyme will have a greater affinity for the transition state than for the substrates.

The reaction coordinates for the uncatalysed and catalysed reactions are shown in Equation 34:

$$E + A + B \overset{K^{*}}{\rightleftharpoons} E + (AB)^{*} \overset{k}{\longrightarrow} E + \text{Products} \qquad \textbf{\textit{uncatalysed}}$$

$$K_s \updownarrow \qquad\qquad K_s^{*} \updownarrow$$

$$E.A.B. \overset{K_E^{*}}{\rightleftharpoons} E(AB)^{*} \overset{k_{cat}}{\longrightarrow} E + \text{Products} \qquad \textbf{\textit{catalysed}}$$

**Note:** Although the enzyme does not take part in the uncatalysed reaction, it has been included to allow direct comparison between the two reactions.

If all binding steps are at equilibrium, $K_s{}^{*}K_E{}^{*} = K^{*}K_s$ and, since the enzyme is a catalyst $k_{cat} >> k$, thus: $K^{*} >> K_E{}^{*}$ (Equation 32) and so $K_s >> K_s{}^{*}$. This indicates that an enzyme must have a much higher affinity for binding the transition state than the native substrate(s); in other words an enzyme facilitates the reaction by binding the transition state much more strongly than the substrate(s).

Antibodies raised to transition state analogues (abzymes) can be quite effective and specific catalysts.

## 5.3 Irreversible inhibitors

### 5.3.1 Non-specific irreversible inhibition

This is the simplest type of irreversible inhibition (see refs 15 and 42); the inhibitor reacts directly with one or more groups on the enzyme to cause inactivation. Inhibition does not involve prior formation of a non-covalent enzyme–inhibitor complex. It can be represented by the simple reaction shown by the equation in Table 1.

The rate of inhibition will depend on the concentrations of enzyme and inhibitor and a second-order rate constant $k$.

$$\frac{d\boldsymbol{e\text{-}i}}{dt} = k\,(\boldsymbol{e} - \boldsymbol{e\text{-}i})(\boldsymbol{i} - \boldsymbol{e\text{-}i}) \qquad [35]$$

where $\boldsymbol{e\text{-}i}$ is the concentration of covalently-inhibited enzyme, $\boldsymbol{e}$ and $\boldsymbol{i}$ are the total concentrations of enzyme and inhibitor, respectively, and $t$ is time. This gives a rather unpleasant second-order equation for the time-course of inhibition, but this can be simplified if concentrations of inhibitor are very much greater than those of the enzyme, which is often the case. Under those conditions the inhibitor concentration remains essentially constant ($(\boldsymbol{i} - \boldsymbol{e\text{-}i}) \approx \boldsymbol{i}$) and the equation from Table 1 can be written as a *pseudo*-first-order reaction:

$$\mathrm{E} \xrightarrow{k.\boldsymbol{i}} \text{E-I and} \frac{d\boldsymbol{e\text{-}i}}{dt} = k.\boldsymbol{i} = k' \qquad [36]$$

where $k' = k.\boldsymbol{i}$ is the *pseudo*-first-order rate constant.

Equation 36 can be integrated to:

$$\ln \boldsymbol{e\text{-}i} = \ln\frac{\boldsymbol{e}_0}{\boldsymbol{e}_t} = 2.303\log\frac{\boldsymbol{e}_0}{\boldsymbol{e}_t} = k't \qquad [37]$$

where $\boldsymbol{e}_o$ and $\boldsymbol{e}_t$ are the enzyme concentrations at the initial (time $t = 0$) and at any subsequent time ($t$), respectively.

If the development of inhibition is followed by measuring the loss of enzyme activity, this relationship may be expressed in terms of the initial enzyme activity ($a_o$) and that remaining at any subsequent time ($a_t$) as the fractional activity remaining:

$$\ln\frac{a_t}{a_o} = -k't \text{ or } \log\frac{a_t}{a_o} = \frac{-k't}{2.303} \qquad [38]$$

Thus a graph of the log of the fractional activity remaining against time will give a straight line with a slope of $-k'$ (see Figure 10a). If a series of such lines is determined at different inhibitor concentrations and a graph of $k'$ is plotted against [I], a straight line passing through the origin with a slope $k$ will be

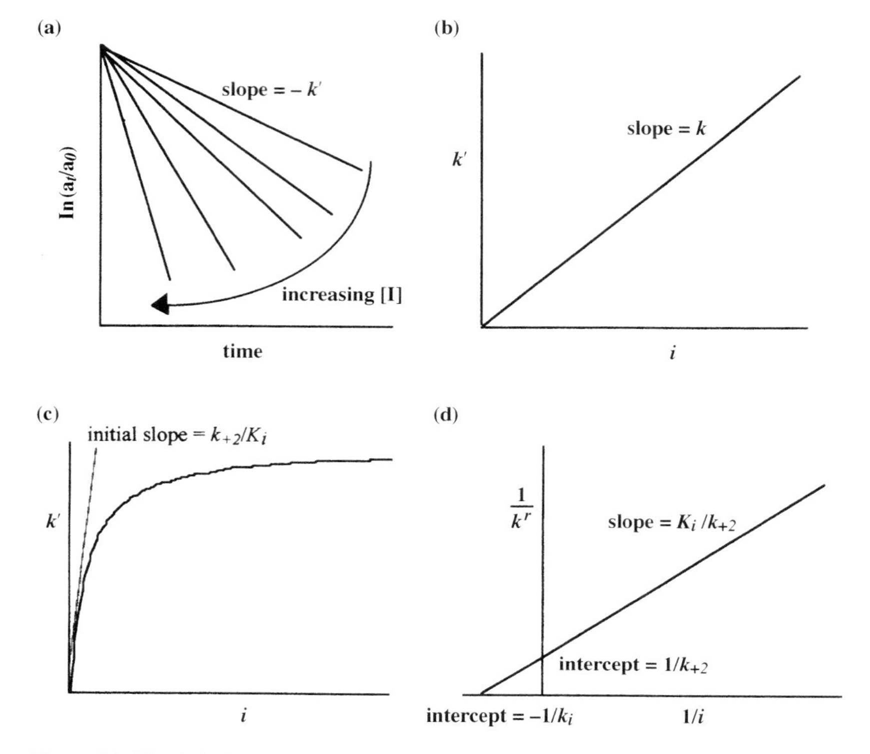


**Figure 10** Kinetic behaviour of irreversible inhibitors. (a) First-order inactivation curves at different inhibitor concentrations. (b) The effects of inhibitor concentration on the apparent first-order rate constant ($k'$) for a non-specific inhibitor. (c and d) The effects of inhibitor concentration on the apparent first-order rate constant ($k'$) for a specific inhibitor.

obtained (see Figure 10b). The value of $k$ is independent of the concentrations of enzyme and inhibitor and has the dimensions of time$^{-1}$ (e.g. s$^{-1}$ or min$^{-1}$), whereas that of $k'$ increases with inhibitor concentration with dimensions of concentration.time$^{-1}$.

The time that it takes one-half of the enzyme activity to be lost ($a_t = a_0/2$), the half-life of the reaction ($t_{1/2}$) will be given by:

$$t_{1/2} = \frac{\ln 2}{k'} = \frac{2.303 \log 2}{k'} = \frac{0.693}{k'} \quad [39]$$

Inhibitors of this type are usually not specific for any particular enzyme but will react with amino acid side chains having similar reactivities in many different enzymes and often with several different residues in the same enzyme. Furthermore most are not entirely specific for a given type of amino acid residue. Some examples are shown in Table 9.

Although this lack of specificity limits their value for many purposes, it can help to identify groups in an enzyme that are necessary for activity.

**Table 9** Some commonly used non-specific enzyme inhibitors

| Reagent/treatment | Amino acid groups affected |
|---|---|
| Acetic anhydride | Acetylates $-NH_2$ groups and also $-OH$ and $-SH$ |
| *N*-Acetylimidazole | Acetylates tyrosine $-OH$ and more slowly with $-NH_2$ |
| 2,3-Butanedione | Reacts with arginine residues |
| 1,2-Cyclohexanedione | Arginine residues; more slowly with $-NH_2$ |
| Diazonium-1-*H*-tetrazole | Histidine and tyrosine; may also react with $-NH_2$ |
| Diethylpyrocarbonate | Histidine; may also react with tyrosine and $-SH$ |
| 2,4-Dinitrofluorobenzene | $-NH_2$, $-SH$, tyrosine $-OH$, and imidazole |
| 5,5′-Dithiobis-(2-nitrobenzoate) | $-SH$ |
| *p*-Hydroxymercuribenzoate (*p*-chloromercuribenzoate) | $-SH$ |
| 2-Hydroxy-5-nitrobenzyl bromide | Reacts with the indole of tryptophan and more slowly with $-SH$ |
| Hydrogen peroxide | Oxidizes methionine and also $-SH$ |
| Iodination | Tyrosine and more slowly histidine |
| Iodoacetate / iodoacetamide | $-SH$; reacts more slowly with tyrosine $-OH$, $-NH_2$, methionine and histidine at lower pH |
| Methylacetimidate | Amino groups |
| *O*-Methylisourea | Amino groups |
| Tetranitromethane | Tyrosine |
| Trinitrobenzene sulfonic acid | Amino groups and also $-SH$ |

**Note**: Some of these reagents are potentially hazardous.
Further details: Wong, S. S. (1991). *Chemistry of protein conjugation and cross-linking.* CRC Press, USA.
Dawson, R. M. C., Elliott, D. C., Elliott, W. H., and Jones, K. M. (1986). *Data for biochemical research.* Oxford Science Publications, UK.

### 5.3.2 Specific irreversible inhibition

This involves the formation of an initial non-covalent enzyme–inhibitor complex (E.I):

$$E + I \underset{k_{-1}}{\overset{k_{+1}}{\rightleftharpoons}} E.I \xrightarrow{k_{+2}} E\text{-}I \qquad [40]$$

which then reacts to give the irreversibly inhibited enzyme (E-I). Thus, the enzyme binds the inhibitor as if were a substrate, forming a complex analogous to the enzyme–substrate complex and reaction then occurs within that complex to give the irreversibly inhibited species. There are two types of specific irreversible inhibitor. In one case the inhibitor contains a chemically reactive group attached to a substrate analogue. Within the enzyme inhibitor complex the high local concentrations of that group and a group on the enzyme surface should lead to a chemical reaction between them resulting in irreversible inhibition. Inhibitors of this type are often known as active site-directed inhibitors (ASDINS). In the second type, the inhibitor is not intrinsically reactive. It first forms a non-covalent complex with the active site of the enzyme and subsequent reaction within that complex leads to the generation of a reactive species that then reacts with the enzyme to form the irreversibly-inhibited species. Inhibitors of this type are known as mechanism-based, enzyme-activated, $k_{cat}$, or suicide inhibitors. They can show a high degree of specificity towards a target enzyme because the generation of the effective inhibitory species from an essentially unreactive compound involves part of the catalytic function of the enzyme itself. Furthermore, the lack of intrinsic reactivity minimizes the possibility of unwanted reactions with other tissue components. Some examples of specific irreversible inhibitors are shown in Table 10.

The mechanism shown in Equation 40 is similar to that of a simple enzyme-catalysed reaction (see Equations 1 and 2) except that the reaction results in a decline of enzyme activity rather than product formation. The rate of irreversible inhibition will be given by a first-order rate constant ($k_{+2}$) and the concentration of the non-covalent E.I complex ($\boldsymbol{e.i}$):

$$\frac{\mathrm{d}\boldsymbol{e\text{-}i}}{\mathrm{d}t} = \mathrm{k}_{+2}\,\boldsymbol{e.i} \qquad [41]$$

As with the Michaelis–Menten relationship, if the inhibitor concentration is much greater than that of the enzyme ($\boldsymbol{i} >> \boldsymbol{e}$) the concentration of the E.I complex will be given by:

$$\boldsymbol{e.i} = \frac{\boldsymbol{e}}{\left(1 + \frac{K_i}{\boldsymbol{i}}\right)} \qquad [42]$$

where $K_i = (k_{-1} + k_{+2}) / k_{+1}$ under steady-state conditions and the simple dissociation constant ($k_{-1} / k_{+1}$) under conditions approximating to thermodynamic equilibrium ($k_{+2} << k_{-1}$).

Combining Equations 41 and 42 gives:

$$\frac{\mathrm{d}\boldsymbol{e\text{-}i}}{\mathrm{d}t} = \frac{k_{+2}\boldsymbol{e}}{\left(1 + \frac{K_i}{\boldsymbol{i}}\right)} \qquad [43]$$

This expression can be integrated to give:

$$k't = \ln\boldsymbol{e} - \ln(\boldsymbol{e} - \boldsymbol{e.i}) \qquad [44]$$

Expressed in terms of decline in enzyme activity, the equation is identical to Equation 38 except that the apparent first-order rate constant ($k'$) is given for Equations 42 and 43 by:

$$k' = \frac{k_{+2}}{\left(1 + \frac{K_i}{\boldsymbol{i}}\right)} \qquad [45]$$

Thus the loss of activity will be first-order with respect to time, as in Figure 10a, but the dependence of the apparent first-order rate constant ($k'$) on inhibitor concentration will be hyperbolic tending to a maximum value of $k_{+2}$ at high inhibitor concentrations and reaching half of this maximum value when $\boldsymbol{i} = K_i$ (see Figure 10c). Thus these two parameters may be determined by fitting the dependence of $k'$ on the inhibitor concentration by non-linear regression, or by

**Table 10** Some specific irreversible inhibitors

| Inhibitor | Enzyme |
|---|---|
| 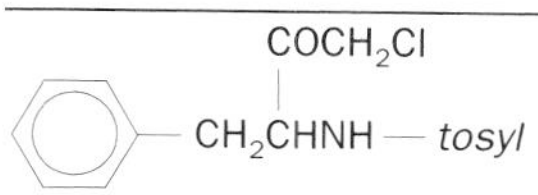 $C_6H_5$—$CH_2CH(COCH_2Cl)NH$—*tosyl*<br>*N*-Tosylphenylalaninechloromethane | Chymotrypsin (EC 3.4.21.2) |
| $H_2NCH_2CH_2CH_2CH_2CH(COCH_2Cl)NH$—*tosyl*<br>*N*-Tosyllysinechloromethane | Trypsin (EC 3.4.21.5) |
| $CH_2=CHCH(NH_2)CH_2CH_2COO^-$<br>γ-vinylGABA | 4-Aminobutyrate transaminase (EC 2.6.1.919) |
| $CH_3OP(=S)(OCH_3)SCH(COOC_2H_5)CH_2COOC_2H_5$<br>Malathion | Acetylcholinesterase (EC 3.1.1.7) |
| 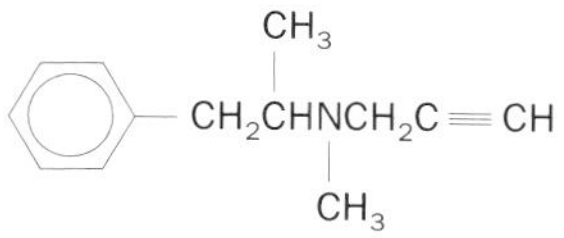 $C_6H_5$—$CH_2CH(CH_3)N(CH_3)CH_2C\equiv CH$<br>Deprenyl (selegiline) | Monoamine oxidase-B (EC 1.4.3.4) |
| 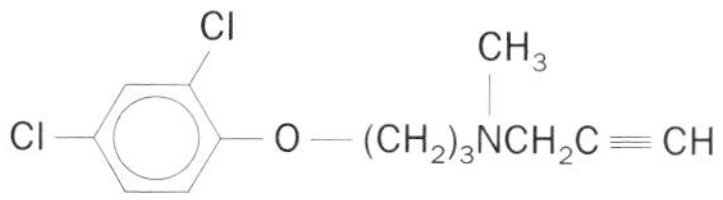 $Cl$—$C_6H_3(Cl)$—$O$—$(CH_2)_3N(CH_3)CH_2C\equiv CH$<br>Clorgyline | Monoamine oxidase-A (EC 1.4.3.4) |
| $^-OOCCH=CHC(CH_3)(NH_3^+)COO^-$<br>α-Methyl-*trans*-dehydroglutamate | Glutamate decarboxylase (EC 4.1.1.15) |
| $CH_2=C=CHCH_2NH(CH_2)_4NHCH_3$<br>*N*-Methyl,*N′*-(2,3-butadienyl)-1,4-butanediamine | Polyamine oxidase (EC 1.5.3.11) |

For further details see references cited in the text and Sandler, M. and Smith, H. J. (ed.) (1989 and 1994). Design of enzyme inhibitors as drugs, Vols 1 and 2. Oxford University Press, UK.

using one of the, less statistically satisfactory, linear transformations of Equation 45. For example in double-reciprocal form the relationship is:

$$\frac{1}{k'} = \frac{K_i}{k_{+2}} \cdot \frac{1}{\boldsymbol{i}} + \frac{1}{k_{+2}} \qquad [46]$$

which will give a linear graph of the form shown in Figure 10d.

A more realistic reaction pathway than that shown in Equation 40 might involve the initial non-covalent complex being transformed into an activated species (E.I*) which subsequently reacts to give the irreversibly-inhibited enzyme:

$$\mathrm{E} + \mathrm{I} \underset{k_{+1}}{\overset{k_{+1}}{\rightleftharpoons}} \mathrm{E.I.} \xrightarrow{k_{+2}} \mathrm{E.I.}^* \xrightarrow{k_{+3}} \mathrm{E\text{-}I} \qquad [47]$$

Steady-state kinetic analysis of such an inhibition mechanism gives the same first-order behaviour as that shown above, except that the kinetic constants are more complex:

$$k' = \frac{\dfrac{k_{+2}.k_{+3}}{(k_{+2} + k_{+3})}}{(1 + K_i/\boldsymbol{i})} \text{ and } K_i = \frac{(k_{-1}.k_{+3} + k_{+2}.k_{+3})}{k_{+1}\,(k_{+2} + k_{+3})} \qquad [48]$$

If $k_{+2} << k_{+3}$ the relationships simplifies to those discussed in relation to Equations 40 (45).

A commonly-used approach to determining the kinetic parameters describing such systems represented by this equation has been the sampling method of Kitz and Wilson (43). In this procedure loss of enzyme activity is determined after increasing times of enzyme–inhibitor pre-incubation and the results are analysed by the procedures illustrated in Figure 10. For such first-order analysis of the time-course of inhibition it is important that the degree of inhibition, measured at each time, represents only irreversible inhibition and that no further time-dependent irreversible inhibition occurs during the period of the assay. These requirements can usually be ensured by a large dilution of the enzyme–inhibitor mixture into the assay and by using high concentrations of assay substrate to ensure displacement of non-covalently bound inhibitor.

Since the mechanism-based inhibition involves the inhibitor binding to the active site of the enzyme, it is often assumed that such inhibition will be strictly competitive with respect to the substrate. However, since most enzyme reactions involve more than one substrate, this may not always be the case, and it is advisable to determine the type of reversible inhibition from separate experiments where the effects of inhibitor concentration on the initial rate of substrate transformation are studied without enzyme–inhibitor pre-incubation. Such experiments will, of course, only be possible where the step leading to irreversible inhibition occurs relatively slowly so that the initial rates of the reaction can be accurately determined. An example of the application of this analysis is the study of the potencies and selectivities of inhibition of the two isoenzymes of monoamine oxidase (EC 1.4.3.4) by the acetylenic inhibitors clorgyline, (−)-deprenyl and pargyline (44) (see also Table 10). That study showed that the relative potencies and for one or other isoenzyme depended on both the affinity for non-covalent binding, as measured by $K_i$, and the rate of reaction in the non-covalent complex to give the irreversibly-inhibited enzyme.

However, this procedure is time-consuming and can suffer considerable errors when the rate of irreversible inhibition is relatively fast. An alternative approach, which is also more appropriate for the study of large numbers of inhibitors, involves the direct determination of reaction progress curves in the presence of substrates and inhibitor. In this procedure the activity of the enzyme is continuously followed by monitoring the appearance of product, or the disappearance of substrate, in assay solutions containing different concentrations of the inhibitor. Under such conditions the reaction velocity will decrease with time as an increasing amount of enzyme becomes irreversibly inhibited, as shown in Figure 11. The overall reaction can be represented by the scheme:

$$\begin{array}{c} E + I \underset{k_{-1}}{\overset{k_{+1}.\boldsymbol{i}}{\rightleftarrows}} E.I \xrightarrow{k_{+3}} E\text{-}I \\ k_{+2}.\boldsymbol{s} \downarrow\uparrow k_{-2} \\ E.S \xrightarrow{k_{cat}} E + \text{Products} \end{array} \quad (49)$$

There have been several approaches to the analysis of such progress curves (45–47). The time-course of product formation in the presence of a mechanism-based inhibitor can be described by the equation:

$$\boldsymbol{p}_t = \boldsymbol{p}_\infty (1 - \exp^{-k'.t}) \quad (50)$$

where $\boldsymbol{p}_t$ and $\boldsymbol{p}_\infty$ are the product concentrations at any time ($t$) and when the enzyme has been completely inhibited, respectively.

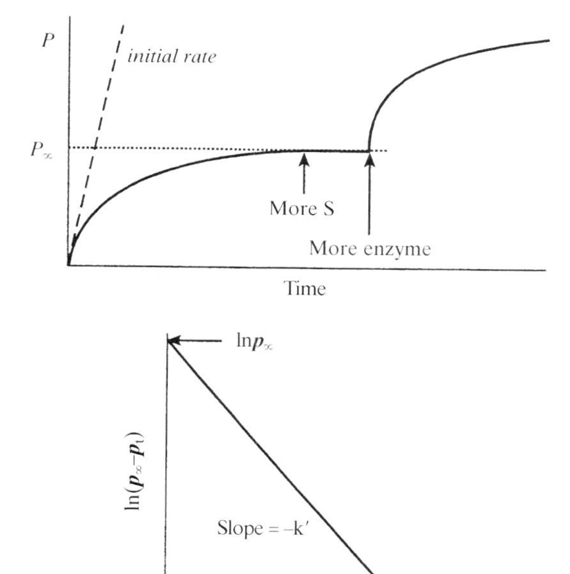


**Figure 11** Progress curve for an enzyme-catalysed reaction in the presence of a fixed amount of an irreversible inhibitor. The lower panel shows the analysis of such a curve for determination of the apparent first-order rate constant for irreversible inhibition.

Equation 50 can be integrated to give:

$$\ln(\boldsymbol{p}_\infty - \boldsymbol{p}_t) = \ln \boldsymbol{p}_\infty - k't \text{ or } \log(\boldsymbol{p}_\infty - \boldsymbol{p}_t) = \log(\boldsymbol{p}_\infty - k't)/2.303 \quad (51)$$

The apparent first-order rate constant, $k'$, will depend on the type of non-covalent interaction with the enzyme. If this is competitive with respect to the substrate, as in Equation 49, the relationship will be:

$$k' = \frac{k_{+3}}{1 + \frac{K_i}{\boldsymbol{i}}\left(1 + \frac{\boldsymbol{s}}{K_m}\right)} \quad [52]$$

where $\boldsymbol{s}$ represents the substrate concentration. At total inhibition, the amount of product formed, $\boldsymbol{p}_\infty$ will be given by (47):

$$\boldsymbol{p}_\infty = K_i.\, k_{cat}.\boldsymbol{s}.\boldsymbol{e} \,/\, K_m.k_{+2}.\boldsymbol{i} \quad [53]$$

Combining Equations 52 and 53 gives:

$$k'.\boldsymbol{p}_\infty = \frac{k_{cat}.\boldsymbol{e}}{1 + \left(1 + \frac{K_i}{\boldsymbol{i}}\right)\left(1 + \frac{\boldsymbol{s}}{K_m}\right)} \quad [54]$$

Thus, a series of reaction progress curves in which the product formed is followed with time at a series of different fixed inhibitor concentrations but all in the presence of the same initial substrate concentration will yield a family of curves which will asymptote at $\boldsymbol{p}_\infty$ and will describe first-order decay curves according to the relationship:

$$\boldsymbol{p}_t = \boldsymbol{p}_\infty\left[1 - \exp^{-}\left\{\frac{k_{cat}.\boldsymbol{e}}{k'.\boldsymbol{p}_\infty}\left(\frac{K_m.k_{+3}.\boldsymbol{i}.t}{K_i.\boldsymbol{s}}\right)\right\}\right] \quad [55]$$

where the apparent first-order rate constant ($k'$) will be given by the exponential term (exp) divided by the time ($t$).

Reaction progress curves, obtained in the presence of different inhibitor concentrations may be fitted to first-order decay curves by direct non-linear regression analysis using one of the many computer programs available (e.g. MacCurveFit or WinCurveFit; see above), in order to evaluate the values of $k'$ and $\boldsymbol{p}_\infty$. If the enzyme concentration (or $V_{max} = k_{cat}.\boldsymbol{e}$) and the $K_m$ value for the substrate are determined separately it is then possible to evaluate the values of $K_i$ and $k_{+3}$ from the relationship given in Equation 54. Such an approach allows the analysis of multiple reaction progress curves, determined for example using a multiple-cuvette spectrophotometer or a microtitre plate reader, by direct data capture and on-line analysis (48). The parameters may also, more laboriously, be determined from plots of ln or log ($\boldsymbol{p}_\infty - \boldsymbol{p}_t$) against time according to Equation 51 (45) or from the time taken for product formation to reach 50% of $\boldsymbol{p}_\infty$ ($t_{1/2}$), since, as shown in Equation 39, $t_{1/2} = 0.693/k'$ (48).

It should be noted that if substrate protects competitively against inhibition by a non-specific irreversible inhibitor (see Section 5.3.1), as depicted in Equation 36, the system:

$$\begin{array}{ccc} E + I & \xrightarrow{k.\boldsymbol{i}} & E\text{-}I \\ k_{+1}.\boldsymbol{s} \downarrow\uparrow k_{-1} & & \\ E.S & \xrightarrow{k_{cat}} & E + \text{Products} \end{array} \qquad [56]$$

will behave in the same way as the specific case (Equations 49–55) except that the apparent first-order rate constant ($k'$) will be given by:

$$k' = \frac{k.\boldsymbol{i}}{1 + \frac{\boldsymbol{s}}{K_m}} \qquad [57]$$

However, there is no *a priori* reason to suppose that substrate will necessarily afford complete protection against a non-specific irreversible inhibitor or, if it does, that the protection will be strictly competitive.

### 5.3.3 Irreversible inhibitors as substrates or substrates as irreversible inhibitors

Since mechanism-based inhibitors behave like substrates in binding to the active site of the enzyme and being converted to the reactive species through a process resembling the normal catalytic process of the enzyme, it is not surprising that it is possible for a proportion of the reactive species to break down to form product. Compounds that behave in this way have sometimes been referred to as 'suicide substrates'. In such cases, the formation of product and the mechanism-based inhibition of the enzyme will be competing reactions according to the following extension of the system shown in Equation 47:

$$E + I \underset{k_{-1}}{\overset{k_{+1}}{\rightleftharpoons}} E.I \xrightarrow{k_{+2}} E.I^* \xrightarrow{k_{+3}} E\text{-}I, \quad E.I^* \xrightarrow{k_{cat}} E + \text{Products} \qquad [58]$$

Time-courses of product formation will be similar to that shown in Figure 11. Provided the inhibitor is present in sufficient excess to cause complete irreversible inhibition of the enzyme the ratio of product formed at complete inhibition ($\boldsymbol{p}_\infty$) divided by the total enzyme concentration ($\boldsymbol{e}_o$) will be a constant, giving the partition ratio ($r$) or the number of mol of product formed per mol of enzyme at complete inhibition:

$$r = \frac{\boldsymbol{p}_\infty}{\boldsymbol{e}_o} = \frac{k_{cat}}{k_{+3}} \qquad [59]$$

The number of mol of inhibitor necessary to inactivate 1 mol of enzyme will thus be given by $(1 + r)$ and so the initial concentration of inhibitor must be greater than $(1 + r)\mathbf{e}_o$ for complete inhibition of the enzyme. If the inhibitor concentration is less than this, the amount of product formed will be given by $r.\mathbf{e}_i$, where $\mathbf{e}_i$ represents the concentration of inhibited enzyme. From this it follows that the amount of inhibited enzyme will be given by:

$$\mathbf{e}_i = \mathbf{e}_o - \frac{\mathbf{i}}{(1 + r)} \text{ or } \frac{\mathbf{e}_i}{\mathbf{e}_o} = 1 - \frac{\mathbf{i}}{(1 + r)\,\mathbf{e}_o} \qquad [60]$$

Therefore, a graph of the fractional activity remaining against the ratio of the initial inhibitor ($\mathbf{i}_o$) concentration to that of the enzyme ($\mathbf{i}_o/\mathbf{e}_o$) will be linear and the value of $(1 + r)$ will correspond to the point where the line intersects the abscissa.

The full equations describing the time-course of product formation in the system represented by Equation 58 are rather cumbersome (see refs 15, 42, 49–51). However, Waley (49) has shown that if the time taken for 50% inhibition, or the time taken for the product concentration to reach 50% of $\mathbf{p}_\infty$ is measured, this half-time ($t_{1/2}$) will correspond to:

$$\mathbf{i}_o.t_{1/2} = \left(\frac{\ln(2 - M)}{1 - M}\right)\frac{K'}{k_{in}} + \frac{\ln 2}{k_{in}}.\mathbf{i}_o \qquad [61]$$

where $k_{in}$ and $K'$ represent the catalytic constant and the apparent Michaelis constant for the inhibitory reaction, respectively, and are given by:

$$k_{in} = \frac{k_{+2}.k_{+3}}{(k_{+2} + k_{+3} + k_{cat})} \text{ and } K' = \left(\frac{k_{-1} + k_{+2}}{k_{+1}}\right)\left(\frac{k_{+3} + k_{cat}}{k_{+2} + k_{+3} + k_{cat})}\right) \qquad [62]$$

and the, dimensionless, constant $M$ is given by:

$$M = \frac{(1 + r)\,\mathbf{e}_o}{\mathbf{i}_o} \qquad [63]$$

Thus, if a series of progress curves is determined at different inhibitor concentrations and under conditions where the ratio $\mathbf{i}_o/\mathbf{e}_o$ is kept constant, a graph of $\mathbf{i}_o.t_{1/2}$ against $\mathbf{i}_o$, will give a straight line with a slope of $0.693/k_{in}$ and the value of $K'$ can be evaluated from the intercept according to Equations 61 and 63. Some examples of compounds that act as both substrates and inhibitors are given in Table 11. Some inhibitors or substrates for monoamine oxidase (EC 1.4.3.4) have, for example, been shown to behave in this way (52, 53). The substrate for dopa decarboxylase (EC 4.1.1.28), L-3,4-dihydroxyphenylalanine (L-dopa) is a time-dependent inhibitor because an occasional transamination reaction results in the conversion of the pyridoxal phosphate coenzyme to the pyridoxamine form during the progress of the assay (see ref. 54). In that case appropriate substitutions on the α-carbon can increase the occurrence of the transamination reaction and give higher partition ratios (see ref. 55).

A rather different case where the inhibitor may, in fact also behave as a substrate for an enzyme occurs when the inhibited species slowly breaks down to yield product(s). In the system:

**Table 11** Some 'suicide substrates'

| Substrate/inhibitor | Enzyme | Reference |
|---|---|---|
| Hydrogen peroxide | Catalase (EC 1.11.1.6) | (a) |
| 2-(*n*-Pentylamino)acetamide (Milacemide) | Monoamine oxidase (EC 1.4.3.4) | (b) |
| 1-Methyl-4-phenyl-1,2,3,6-tetrahydropyridine (MPTP) | Monoamine oxidase (EC 1.4.3.4) | (c) |
| *S*-Adenosylmethionine | 1-Aminocyclopropane-1-carboxylate synthase (EC 4.4.1.14) | (d) |
| Dopa | Dopa decarboxylase (EC 4.1.1.28) | (e) |
| α-Methyldopa | Dopa decarboxylase (EC 4.1.1.28) | (f) |
| Serpins | Sperm peptidase acrosin (EC 3.4.21.10) | (g) |
| Plasminogen-activator inhibitor 1 | Thrombin (EC 3.4.21.6) | (h) |

References:

(a) Ghadermarzi, M. and Moosavi-Movahedi, A. A. (1999). *Biochim. Biophys. Acta*, **1431**, 30.
(b) O'Brien, E. M., McCrodden, J. M., Youdim, M. B. H., and Tipton, K. F. (1994). *Biochem. Pharmacol.*, **47**, 617.
(c) Tipton, K. F., McCrodden, J. M., and Youdim, M. B. H. (1986). *Biochem. J.*, **240**, 379.
(d) Casas, J. L., Garcia-Canovas, F., Tudela, J., and Acosta, M. (1993). *J. Enzyme Inhib.*, **7**, 1.
(e) Minelli, A., Charteris, A. T., Voltattorni, C. B., and John, R. A. (1979). *Biochem. J.*, **183**, 361.
(f) Palfreyman, M. G., Bey, P., and Sjoerdama, A. (1987). In *Essays in biochemistry* (ed. R. D. Marshall and K. F. Tipton), Vol. 23, p. 28. Academic Press Ltd, London.
(g) Engh, R. A., Huber, R., Bode, W., and Schulze, A. J. (1995). *Trends Biotechnol.*, **13**, 503.
(h) van Meijer, M., Smilde, A., Tans, G., Nesheim, M. E., Pannekoek, H., and Horrevoets, A. J. (1997). *Blood*, **90**, 1874.

$$E + 1 \underset{k_{-1}}{\overset{k_{+1}}{\rightleftharpoons}} E.I \xrightarrow{k_{+2}} E\text{-}I \xrightarrow{k_{cat}} E + \text{Product(s)} \qquad [64]$$

inhibition may appear irreversible if $k_{cat}$ is very small and product release is so slow that it does not occur to any significant extent during the time-course of inhibition. Several inhibitors of acetylcholinesterase (EC 3.1.1.7), for example, behave in this way (56, 57).

In such cases the inhibition kinetics may be studied by the approaches discussed in connection with Equations 40 and 47. The constant $k_{cat}$ can then be determined by measuring the rate of recovery of fully-inhibited enzyme after removal of excess free inhibitor, for example, by gel filtration or dilution according to the relationship:

$$\ln (a_{\infty} - at) = K_{cat}.t + constant \qquad [65]$$

where $a_{\infty}$ and $a_t$ are the activities after complete recovery of activity and at any intermediate time, $t$, respectively

Inhibition may also appear to be irreversible if the inhibitor binds tightly, but non-covalently, to the enzyme and the dissociation is so slow that it cannot be detected within the time-scale of the experiment, as may be the case with transition state analogues (Section 5.2). It would be necessary to assay for product formation in order to distinguish between tight-binding inhibitors and very poor substrates.

# 6 Behaviour *in vivo*

## 6.1 Recovery from irreversible inhibition

After *in vivo* treatment with a single dose of a reversible inhibitor, the rate of recovery of enzyme activity will depend of the rate of removal of the inhibitor from the tissues by metabolism and elimination. For an irreversible inhibitor the recovery requires synthesis of new enzyme to replace the inhibited enzyme. After inhibition the enzyme activity will recover with time as new enzyme is synthesized (Figure 12). The kinetics of recovery will be complex during any time where free irreversible inhibitor remains in the tissue, since this may inhibit any newly synthesized enzyme. However, after all free inhibitor has been removed from the tissues the analysis of recovery is simple. The turnover of enzyme molecules is of course a dynamic process and the amount of enzyme molecules in the cell at any at time will be a balance between it rates of synthesis and removal. This can be represented by the simplified model:

$$\text{Precursors} \xrightarrow{k_{synthesis}} \text{Enzyme} \xrightarrow{k_{degradation}} \text{Breakdown products} \qquad [66]$$

in which the complex processes resulting in the synthesis of new enzyme are countered by its breakdown. Thus the reappearance of active enzyme can be represented by the equation:

$$\frac{d\boldsymbol{e}}{dt} = k_{synthesis} \cdot [\text{Precursors}] - k_{degradation} \cdot \boldsymbol{e} \qquad [67]$$

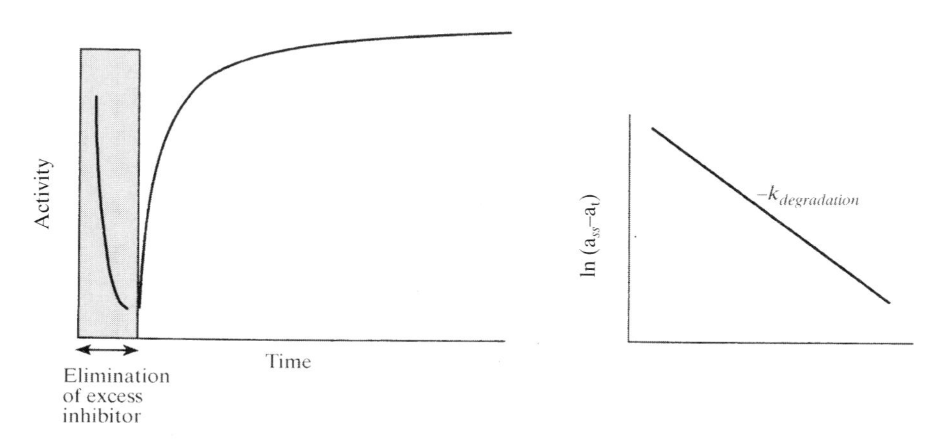


**Figure 12** Recovery of enzyme activity *in vivo* after administration of a single dose of an irreversible inhibitor.

Eventually the recovery of active enzyme will level off at a constant, steady-state, level where the rate of synthesis is exactly counterbalanced by the rate of degradation:

$$k_{synthesis} \cdot [\text{Precursors}] = k_{degradation} \cdot \boldsymbol{e}_{\boldsymbol{ss}} \quad [68]$$

where $\boldsymbol{e}_{ss}$ is the steady-state enzyme concentration. Equation 68 may be used to eliminate the rate of synthesis, which depends on a number of different components, from Equation 67 giving:

$$\frac{\mathrm{d}\boldsymbol{e}}{\mathrm{d}t} = k_{degradation} \cdot (\boldsymbol{e}_{\boldsymbol{ss}} - \boldsymbol{e}) \quad [69]$$

This is a first-order process which integrates to give (see Equations 36–38):

$$\ln(\boldsymbol{e}_{\boldsymbol{ss}} - \boldsymbol{e}) = k_{degradation} \cdot t + \text{constant} \quad [70]$$

If the recovery of enzyme activity is determined, this relationship can be written as:

$$\ln \frac{a_{ss}}{a_t} = k_{degradation} \cdot t + \text{constant} \quad [71]$$

where $a_t$ is the activity at any time ($t$), and $a_{ss}$ is the final, constant, steady-state enzyme activity. Thus the apparent first-order rate constant can be determined as shown in Figure 12 (see also Equations 36–38) and the half-life for enzyme turnover will be $0.693/k_{degradation}$ (see Equation 39). The rates of turnover of different cellular enzymes vary widely, as shown for some rat liver enzymes in Table 12. The rates of turnover of an enzyme can also differ greatly between tissues and species as shown for monoamine oxidases A and B in Figure 13. This can have important implications for therapy with irreversible inhibitors.

As discussed above, the rates of recovery from the effects of a reversible inhibitor will depend on its rate of removal from the tissues by metabolism and elimination (see e.g. ref. 58). Metabolic differences can result in considerable variations in drug clearance. A particularly striking example of this is seen with the benzodiazepines (59). For comparison with the data in Table 12, the rate of recovery from a single dose of several reversible monoamine oxidase inhibitors (administered *p.o.* or *i.p.*) in the human and in experimental animals is about 12–18 hours (see e.g. ref. 60).

## 6.2 Effects of cellular processes

Assuming that it penetrates the tissues, the effects on an excess of an irreversible inhibitor will be the complete cessation of metabolic processes that depend on the activity of the affected enzyme. The effects on a metabolic pathway of administration of an amount of irreversible inhibitor that causes only partial inhibition of an enzyme will depend on the degree of flux control exerted by that enzyme. In some cases it may be necessary to cause substantial inhibition of the activity of an individual enzyme before any significant effects on metabolic flux are seen (see e.g. refs 61 and 62). Similar conclusions also apply to the rate of recovery of the process after a single dose of a reversible inhibitor. Where the

**Table 12** The rates of turnover of some enzymes in rat liver

| Enzyme | Half-life (hour) |
|---|---|
| Ornithine decarboxylase (EC 4.1.1.17) | 0.2 |
| 5-Aminolaevulinate synthase (EC 2.3.1.37 | 1.2 |
| DNA-directed RNA polymerase (EC 2.7.7.6) | 1.3 |
| Tyrosine aminotransferase (EC 2.6.1.5) | 1.5 |
| Tryptophan 2,3-dioxygenase (EC 1.13.11.11) | 2..0 |
| Hydroxymethylglutaryl-CoA reductase (EC 1.1.1.88) | 4.0 |
| Phospho*enol*pyruvate carboxykinase (EC 4.1.1.49) | 12 |
| DNA-directed RNA polymerase (EC 2.7.7.6) | 13 |
| Glucose 6-phosphate dehydrogenase (EC 1.1.1.49) | 24 |
| Glucokinase (EC 2.7.1.12) | 30 |
| Catalase (EC 1.11.1.6) | 33 |
| Acetyl-CoA carboxylase (EC 6.4.1.2) | 50 |
| Glyceraldehyde 3-phosphate dehydrogenase (EC 1.2.1.12) | 74 |
| Pyruvate kinase (EC 2.7.1.40) | 84 |
| Arginase (EC 3.5.3.1) | 108 |
| Fructose-bisphosphate aldolase (EC 4.1.2.13) | 117 |
| Lactate dehydrogenase (EC 1.1.1.27) | 144 |
| 6-Phosphofructokinase (EC 2.7.1.11) | 168 |

Data taken from:
Schimke, R. T. (1973). *Adv. Enzymol.*, **37**, 137.
Goldberg, A. L. and Dice, J. F. (1974). *Annu. Rev. Biochem.*, **43**, 835.
Schimke, R. T. and Katanuma, N. (1975). *Intracellular protein turnover*, pp. 131–283. Academic Press, New York.
Hopgood, M. F. and Ballard, F. J. (1974). *Biochem. J.*, **144**, 371.

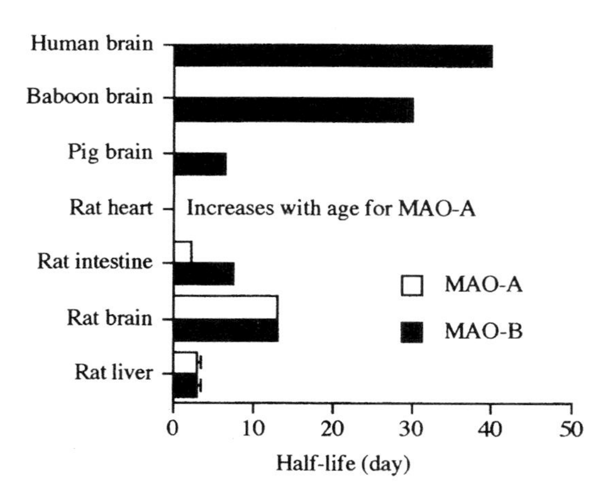


**Figure 13** Tissue and species differences in the half-lives of the monoamine oxidase (EC 1.4.3.4) isoenzymes A and B.

enzyme involved shows little flux control the recovery may take a considerably shorter time than that required for complete recovery of enzyme activity. Similarly the pharmacological effects may have a relatively shorter duration. For example, monoamine oxidase B inhibition enhances the behavioural response to injected 2-phenylethylamine but this effect is lost when only about 20% of the enzyme activity has recovered (63).

The behaviour of reversible inhibitors will depend in the type of inhibition involved. Any type of inhibitor would be expected to result in an increase in the steady-state levels of the substrate(s) for the enzyme and a corresponding decrease in the amount of product(s). In the case of a non-competitive inhibitor the increased substrate concentration will have no effect on the degree of inhibition of the enzyme. However, increasing substrate concentrations would be expected to reduce the effects of a competitive inhibitor but actually enhance the effects of an uncompetitive inhibitor (see Figure 4). The effects the inhibitors have on the levels of product(s) will depend on whether their steady-state levels are sufficient to give significant inhibition of the enzyme and on the type of product inhibition involved (see Table 7). However, clearly if the 'normal' steady-state product concentrations were sufficient to cause significant inhibition of the enzyme, a reduction of product concentration as a result of administration of a reversible inhibitor would be expected to result in a relief of this inhibition.

## 6.3 Inhibitor specificity

Many enzyme inhibitors have proven useful as therapeutic or prophylactic argents, some examples are shown in Table 13. However, although it is often assumed that enzyme inhibitors are highly specific towards a single target enzyme, closer examination reveals that rather few inhibitors have such a high degree of selectivity. Many reversible and irreversible inhibitors are known to affect more than one enzyme. For example, the nitric oxide synthase (EC 1.14.13.39) inhibitors diphenyleneiodonium and monoaminoguanidine also inhibit NADH:ubiquinone oxidoreductase (EC 1.6.5.3) and both aldose reductase (EC 1.1.1.21) and the copper-containing amine oxidases, respectively (64–66). The acetylcholinesterase (EC 3.1.1.7) inhibitor DFP (diisopropylphosphoroluoridate) is known to inhibit a large family of serine peptidases, including chymotrypsin (EC 3.4.21.2), thrombin (EC 3.4.21.6), and trypsin (EC 3.4.21.5) (see ref. 1). The γ-aminobutyrate transaminase (GABA transaminase; EC 2.6.1.19) inhibitors 4-aminohex-5-ynoate and 4-aminohex-5-enoate also inhibit ornithine transaminase (EC 2.6.1.13) and aspartate transaminase (EC 2.6.1.1) (67).

Even with an inhibitor that appears to have a very high specificity towards a single enzyme, the possibility that metabolites formed *in vivo* may affect other systems cannot be excluded. The phenomenon of lethal synthesis, where, for example, flouroacetate is metabolized to the potent aconitase (EC 4.2.1.3) inhibitor fluorocitrate, is well-known (see e.g. ref. 68) and it is not surprising that metabolism may result in a change in the activities of an otherwise specific inhibitor. For example, the compound *l*-deprenyl (selegiline) is highly specific as an inhibitor of the monoamine oxidases, although it does stimulate growth factor

**Table 13** Some clinically relevant enzyme inhibitors

| Inhibitor | Enzyme | Application |
|---|---|---|
| Asprin | Prostaglandin synthase (EC 1.14.99.1) | Anti-inflammatory |
| Allopurinol | Xanthine dehydrogenase (EC 1.1.1.204) | Gout |
| $\epsilon$-Aminocaproate | Plasmin (EC 3.4.21.7) | Antifibrinolytic |
| Azaserine | Phosphoribosylformyl-glycinamidine synthase (EC 6.3.5.3) | Anticancer |
| Benserazide; carbodopa | Dopa decarboxylase (peripheral) (EC 4.1.1.28) | Parkinson's disease |
| Captopril | Peptidyl-dipeptidase A (EC 3.4.15.1) | Antihypertensive |
| Clorgyline | Monoamine oxidase-A (EC 1.4.3.4) | Antidepressant |
| Clavulanic acid | $\beta$-Lactamase (EC 3.5.2.6) | Antibiotic resistance |
| D-Cycloserine | Alanine racemase (EC 5.1.1.1) | Antibacterial |
| Deprenyl (selegiline) | Monoamine oxidase-B (EC 1.4.3.4) | Parkinson's disease |
| Disulfiram | Aldehyde dehydrogenase (EC 1.2.1.3) | Alcoholism |
| 5-Fluorouracil | Thymidylate synthase (EC 2.1.1.45) | Cancer chemotherapy |
| Lithium ion | Inositol-1,4-bisphosphate 1-phosphatase (EC 3.1.3.57) | Mania/depression |
| Neostigmine | Acetylcholinesterase (EC 3.1.1.7) | Myesthenia gravis; glaucoma |
| Omeprazole | $H^+/K^+$-exchanging ATPase (EC 3.6.3.10) | Anti-ulcer |
| Penicillins | D-Alanyl-D-alanine carboxypeptidase (EC 3.4.16.4) | Antibiotics |
| Sildenafil (Viagra) | 3′,5′-cyclic-GMP phosphodiesterase (EC 3.1.4.35) | Erectile problems |
| Sulfonamides | Dihydropteroate synthase (EC 2.5.1.15) | Antibacterial |
| Tacrine | Acetylcholinesterase (EC 3.1.1.7) | Alzheimer's disease |
| Valproate | Aldehyde reductase (EC 1.1.1.19 and EC 1.1.1.2) | Anticonvulsant |
| $\gamma$-VinylGABA | 4-Aminobutyrate transaminase (EC 2.6.1.19) | Anticonvulsant |

production by an independent mechanism and is converted to *l*-amphetamine that inhibits presynaptic amine transport by the action of members of the cytochrome P-450-dependent hydroxylase family of enzymes (see ref. 69). Such considerations mean that it is ill-advised to assume that the effects seen following administration of an enzyme inhibitor result solely from inhibition of that enzyme unless the specificity of the inhibitor, and any metabolites that may be formed from it, has been rigorously established.

## References

1. International Union of Biochemistry and Molecular Biology: Enzyme Nomenclature. http://www.chem.qmw.ac.uk/iubmb/enzyme/
2. Tipton, K. F. and Boyce, S. (2000). In *Encyclopedia of life sciences*. Macmillan Reference Co. UK. Article 710, http://www.els.net

3. Colowick, S. P. and Kaplan, N. O., and others (1955ff). *Methods in enzymology*, Vols 1–. Academic Press Inc., New York.
4. Bergmeyer, H.-U. (1983ff). *Methods of enzymatic analysis*, 3rd edn, Vols 1–10. Verlag Chemie, Weinheim.
5. Barman, T. E. (1969 and 1974). *Enzyme handbook*. Springer–Verlag, Heidelberg.
6. Tipton, K. F. (1992). In *Enzyme assay: a practical approach* (ed. R. Eisenthal and M. J. Danson), pp. 1–58. IRL Press, Oxford.
7. Nicholls, R. G., Jerfy, M., and Roy, A. B. (1974). *Anal. Biochem.*, **61**, 93.
8. MacCurveFit and WinCurveFit, Kevin Raner Software, Australia – http://www.home.aone.net.au/krs
9. Henderson, P. J. F. (1992). In: *Enzyme assay: a practical approach* (ed. R. Eisenthal and M. J. Danson), pp. 277–316. IRL Press, Oxford.
10. Tipton, K. F. (ed.) (1985). In *Techniques in the life sciences*, Vol. B1/II, Supplement p. B5 113/1. Elsevier Ltd., Ireland.
11. Whitehead, E. P. (1970). *Prog. Biophys. Mol. Biol.*, **21**, 323.
12. Tipton, K. F. (1979). In *Companion to biochemistry*, Vol. 2 (ed. A. T. Bull, J. R. Lagnado, J. O. Thomas, and K. F. Tipton), pp. 327–82. Longman, London.
13. Hill, A. V. (1910). *J. Physiol.* (London), **40**, 4.
14. Cornish-Bowden, A. (1995). *Fundamentals of enzyme kinetics*. Portland Press, London.
15. Tipton, K. F. (1996). In *Enzymology LabFax* (ed. P. C. Engel), pp. 115–74. Bios Scientific Publishers, Oxford, UK, & Academic Press, San Diego, USA.
16. Segel, I. H. (1993). *Enzyme kinetics: behavior and analysis of rapid equilibrium and steady-state enzyme systems*. Wiley, UK.
17. Bardsley, W. G. and Childs, R. E. (1975). *Biochem. J.*, **149**, 313.
18. Teipel, J. and Koshland, D. E. (1989). *Biochemistry*, **8**, 4656.
19. Engel, P. C. and Ferdinand, W. P. C. (1973). *Biochem. J.*, **131**, 97.
20. Cornish-Bowden, A. (1988). *Biochem. J.*, **250**, 309.
21. Dixon, H. B. F. and Tipton, K. F. (1973). *Biochem. J.*, **133**, 837.
22. Spears, G., Sneyd, J. T., and Loten, E. G. (1971). *Biochem. J.*, **125**, 1149.
23. Zollner, H. Z. (1999). *Handbook of enzyme inhibitors*. John Wiley, UK.
24. Dixon, M. (1953). *Biochem. J.*, **55**, 170.
25. Dixon, M. and Webb, E. C. (1979). *Enzymes*. Longman, London.
26. Cleland, W. W. (1989). *Biochim. Biophys. Acta*, **1000**, 209.
27. Fromm, H. J. (1979). In *Methods in enzymology*, Vol. 63 (ed. D. L. Purich), p. 42. Academic Press, New York.
28. Rudolph, F. B. (1979). In *Methods in enzymology*, Vol. 63 (ed. D. L. Purich), p. 411. Academic Press, New York.
29. Dalziel, K. (1969). *Biochem. J.*, **114**, 547.
30. Cleland, W. W. (1989). In *The enzymes* (ed. P. D. Boyer), p. 1. Academic Press, New York.
31. Elliott, K. R. F. and Tipton, K. F. (1974). *Biochem. J.*, **141**, 789.
32. Huang, C. Y. (1979). In *Methods in enzymology*, Vol. 63 (ed. D. L. Purich), p. 486. Academic Press, New York.
33. Dewolf, W. and Segel, I. H. (2000). *J. Enzyme Inhib.*, **15**, 311.
34. Cleland, W. W. (1979). In *Methods in enzymology*, Vol. 63 (ed. D. L. Purich), p. 500. Academic Press, New York.
35. Warren, G. B. and Tipton, K. F. (1974). *Biochem. J.*, **139**, 311 and 321.
36. Morrison, J. F. (1979). In *Methods in enzymology*, Vol. 63 (ed. D. L. Purich), p. 257. Academic Press, New York.
37. Storer, A. C. and Cornish-Bowden, A. (1976). *Biochem. J.*, **159**, 1.
38. Williams, J. W. and Morrison, J. F. (1979). In *Methods in enzymology*, Vol. 63 (ed. D. L. Purich), p. 437. Academic press, New York.

39. Henderson, P. J. F. (1973). *Biochem. J.*, **135**, 101.
40. Dixon, M. (1972). *Biochem. J.*, **129**, 197.
41. Radzicka, A. and Wolfenden, R. (1995). In *Methods in enzymology*, Vol. 249 (ed. D. L. Purich), p. 284. Academic Press, New York.
42. Tipton, K. F. (1989). In *Design of enzyme inhibitors as drugs* (ed. M. Sandler and H. J. Smith), pp. 70–93. Oxford University Press, London.
43. Kitz, R. and Wilson, I. B. (1962). *J. Biol. Chem.*, **337**, 3245.
44. Fowler, C. J., Mantle, T. J., and Tipton, K. F. (1982). *Biochem. Pharmacol.*, **31**, 3555.
45. Liu, W. and Tsou, C. L. (1986). *Biochim. Biophys. Acta*, **242**, 185.
46. Forsberg, A. and Puu, G. (1984). *Eur. J. Biochem.*, **140**, 153.
47. Walker, B. and Ellmore, D. T. (1984). *Biochem. J.*, **221**, 277.
48. Tipton, K. F., Balsa, D., and Unzeta, M. (1993). In *Monoamine oxidase: basic and clinical aspects* (ed. H. Yashuhara, S. H. Parvez, K. Oguchi, M. Sandler, and T. Nagatsu), pp. 1–13. VSP BV, Utrecht, The Netherlands.
49. Waley, S. G. (1980). *Biochem. J.*, **185**, 771.
50. Waley, S. G. (1985). *Biochem. J.*, **277**, 843.
51. Tatsunami, S., Yago, M., and Hosoe, M. (1981). *Biochim. Biophys. Acta*, **662**, 226.
52. Tipton, K. F., Fowler, C. J., McCrodden, J. M., and Strolin Benedetti, M. (1983). *Biochem. J.*, **209**, 235.
53. Tipton, K. F., McCrodden, J. M., and Youdim, M. B. H. (1986). *Biochem. J.*, **240**, 379.
54. Minelli, A., Charteris, A. T., Voltattorni, C. B., and John, R. A. (1979). *Biochem. J.*, **183**, 361.
55. Palfreyman, M. G., Bey, P., and Sjoerdama, A. (1987). In *Essays in biochemistry* (ed. R. D. Marshall and K. F. Tipton), Vol. 23, p. 28. Academic Press Ltd., London.
56. O'Brien, R. D. (1968). *Mol. Pharmacol.*, **4**, 121.
57. Aldridge, W. N. (1989). In *Design of enzyme inhibitors as drugs* (ed. M. Sandler and H. J. Smith), pp. 294–313. Oxford University Press, London.
58. Greenblatt, D. J. (1993). *J. Clin. Psychiatry*, **54** (suppl.) 8.
59. Carlile, D. J., Zomorodi, K., and Houston, J. B. (1997). *Drug Metab. Dispos.*, **25**, 903.
60. Waldmeier, P. C., Felner, A. E., and Tipton, K. F. (1983). *Eur. J. Pharmacol.*, **94**, 73.
61. Fell, D. (1996). *Understanding the control of metabolism*. Portland Press, UK.
62. Bakker, B. M., Westerhoff, H. V., Opperdoes, F. R., and Michels, P. A. (2000). *Mol. Biochem. Parasitol.*, **106**, 1.
63. Youdim, M. B. H. and Tipton, K. F. (2002). *Parkinsonism Rel. Dis.*, **8**, 247.
64. Majander, A., Finel, M., and Wikstrom, M. (1994). *J. Biol. Chem.*, **19**, 21037.
65. Kumari, K., Umar, S., Bansal, V., and Sahib, M. K. (1991). *Biochem. Pharmacol.*, **41**, 1527.
66. Yu, P. H. and Zuo, D. M. (1977). *Diabetologia*, **40**, 1243.
67. John, R. A., Jones, E. D., and Fowler, L. J. (1979). *Biochem. J.*, **177**, 721.
68. Clarke, D. D. (1991). *Neurochem. Res.*, **16**, 1055.
69. Tipton, K. F. (1994). *Clin. Pharmacol. Ther.*, **56**, 781.

# Chapter 2
# Gene expression

Luis Parada, Cem Elbi, Miroslav Dundr, and Tom Misteli
National Cancer Institute, NIH, Bethesda, MD 20892, USA.

## 1 Introduction

Gene expression has traditionally been studied by biochemical, molecular, and genetic approaches, rather than cell biological methods. Recent advances in microscopy techniques and microscope equipment, probe generation, and the availability of virtually any sequence in the genome as the result of sequencing efforts have made cell biological investigation of gene expression and genome organization a desirable and feasible undertaking (1, 2). The cell biological study of gene expression almost invariably involves microscopy methods. It is now possible to routinely and reliably localize nuclear proteins, entire chromosomes, specific gene loci, as well as unspliced pre-mRNA and spliced mRNA in chemically-fixed cells. Combination of these methods and application of image analysis software provide powerful approaches to the cell biological study of gene expression.

The most basic information about expression of genes comes from the localization of particular nuclear proteins by indirect immunofluorescence microscopy using specific antibodies (Protocol 1). This method is well-established, reliable, and quick if an antibody against the protein of interest is available. However, the discovery of autofluorescent proteins that can be expressed as fusion proteins with any available cDNA is rapidly replacing the antibody method (Protocols 2 and 3). Proteins within the nucleus are often insoluble or only partially soluble. Sequential extraction of nuclei ultimately resulting in an insoluble fraction termed the nuclear matrix fraction can be used to probe the extractability of nuclear proteins (Protocol 4).

Nascent RNA, and therefore as a first approximation active transcription sites, can be visualized by incorporation of bromo-labelled UTP into cells (Protocols 5 and 6). For the more specific detection of particular genes or RNAs *in situ* hybridization is used (Protocols 7–11). This method is based on the hybridization of a labelled, specific probe to a known DNA or RNA sequence. Traditionally probes consisted of nick translated fragments (Protocol 7). Recently the use of riboprobes generated by *in vitro* transcription, or specifically synthesized oligonucleotide probes has dramatically reduced the size of the required target sequence (Protocol 8). Although, in principle rather straightforward, *in situ* hybridization can be challenging depending on the target and probe selected. Optimization of standard protocols such as the ones described in this chapter (Protocols 9–11) is often essential.

Entire chromosomes can be detected either in metaphase or interphase cells by chromosome painting (Protocols 12–16). This method is a derivative of the *in situ* hybridization protocol with the difference that probes are generated against an entire chromosome rather than a specific locus. Chromosome painting is the basis for many modern cytogenetic methods such as multi-colour FISH and spectral karyotyping in which each chromosome is painted a different colour in a single experiment. These methods are becoming widely popular in cytogenetic analyses of chromosomal abnormalities and are beyond the scope of this chapter (for a review see ref. 3).

Many types of detection methods involving various colorimetric substrates are available for each of these methods. By far the most reliable, sensitive, and reproducible is detection of fluorescently tagged probes. All protocols in this chapter describe fluorescence probes, but if desired, any colorimetric method can be used.

## 2 Indirect immunofluorescence

Indirect immunofluorescence (IF) is the method of choice when attempting to localize a protein in a cell. The method is based on the specific interaction of an antibody with its target inside a chemically-fixed, permeabilized cell. The immobilized antibody is visualized in a detection reaction using a fluorescently tagged secondary antibody specifically recognizing the epitope-specific primary antibody.

IF has been widely used to detect proteins in virtually all cellular locations. Crucial for a successful IF experiment is the availability of a specific antibody. It is essential that potential cross-reactivities of any antibody used in IF experiments is eliminated by testing the antibody by Western blotting. Never use an antibody that gives more than one band on a Western blot! Note that antibodies, which work well on a Western blot, do not necessary give good fluorescence microscopy signals. It is further essential that the signal-to-noise ratio of the detection antibody is as high as possible. All secondary antibodies give some background. Nowadays background problems are rare since highly purified secondary antibodies are commercially available. The cell nucleus contains a vast amount of highly charged molecules, particularly DNA and RNA, which can cause some antibodies to stick non-specifically. If desired, although we routinely do not use it, 0.1% fish skin gelatin or 1% normal goat serum can be included in all wash and incubation buffers to reduce background.

A further essential point is fixation. We recommend the use of paraformaldehyde fixation. Paraformaldehyde covalently crosslinks proteins. Like all fixation methods, paraformaldehyde does not completely preserve cellular morphology, but it gives superior results compared to most other fixation methods. Note that a separate permeabilization step is required when paraformaldehyde fixation is used. Some antibodies do not work in paraformaldehyde fixation. The most practical alternative is the use of methanol fixation, which results in permeabilization of cells by extraction of lipids from the membranes and the precipitation of cellular proteins. Methanol fixation is harsh, distorts the cellular morphology significantly, and should only be used if other methods fail. A useful compromise that gives better structural preservation than methanol fixation but is harsher is the use of 3.7% paraformaldehyde in 5% acetic acid (4).

## Protocol 1

## Indirect immunofluorescence

### Equipment and reagents

- Glass coverslips, 22 × 22 mm
- 6-well plates
- Forceps, Dumont GG (Electron Microscopy Sciences)
- Microscopy coverslide
- Parafilm
- PBS pH 7.4
- Paraformaldehyde (granular; Electron Microscopy Sciences)
- Fluorescently labelled secondary antibody (Vector Laboratories)
- Mounting medium (Molecular Probes)

### Method

1. Prepare sterile 22 × 22 mm No. 1.5 glass coverslips by soaking them in 95% ethanol[a] and gently flaming them. For multiple samples use 6-well plates.[b]
2. Grow cells on coverslips to 50–70% confluence.[c]
3. Fix cells in 3–4 ml of 2% formaldehyde in PBS pH 7.4 for 15 min at room temperature.[d,e]
4. Quickly rinse cells in 4 ml PBS, and wash the coverslips in 2 ml PBS, twice for 5 min at room temperature.[f]
5. Permeabilize cells by removing the wash solution and adding 2 ml of 0.5% Triton X-100 in PBS, for 5 min on ice.
6. Quickly rinse cells in 4 ml PBS, and wash the coverslips in 2 ml PBS, three times for 8 min at room temperature.
7. Place a piece of Parafilm large enough to easily accommodate all coverslips on a flat surface.
8. For each coverslip, place 50 μl of the appropriate concentration of primary antibody on the Parafilm. Make sure to separate the drops generously.
9. Remove the coverslip from the 6-well plate using fine forceps.[g] Remove excess PBS by blotting the edges of the coverslip gently against a piece of filter paper.
10. Invert the coverslip with the cells facing down onto the antibody drop.

**Protocol 1** continued

**11** Incubate the coverslips for 1 h at room temperature inverted onto a piece of Parafilm. Cover the Parafilm with a light-tight lid.

**12** Remove the coverslip from the Parafilm by squirting 1 ml of PBS under the coverslip to float it up. Place the coverslip with the cells facing up in the 6-well plate.

**13** Wash the coverslip in 2 ml PBS, three times for 8 min at room temperature.

**14** Repeat steps 6–9 with a drop of appropriately diluted secondary antibody (2–4 μg/ml in PBS).

**15** Incubate the coverslip for 1 h at room temperature inverted onto a piece of Parafilm.

**16** Wash the coverslip in 2 ml PBS, three times for 8 min at room temperature.[h]

**17** Place 10 μl of mounting medium on a glass microscopy slide. Remove the coverslip from the 6-well plate. Remove excess PBS by blotting the edges of the coverslip gently against a piece of filter paper.

**18** Invert the coverslip cells facing down onto the mounting medium.[i]

**19** Remove excess mounting medium by holding two pieces of filter paper against two opposing edges of the coverslip.[j]

[a] It is most convenient to store a large number of coverslips in 95% ethanol.

[b] All volumes given are optimized for 6-well plates and 22 × 22 mm coverslips.

[c] Confluent lawns of cells result in increased background signal.

[d] 2% formaldehyde is made up fresh prior to use by dissolving the appropriate amount of EM grade paraformaldehyde in PBS in a Pyrex bottle on a hot plate with a stir bar. Paraformaldehyde goes into solution at 70 °C. Do not tighten the bottle cap so as to avoid build-up of pressure. Place the dissolved solution on ice to cool down and check pH.

[e] Alternative fixation procedures include methanol at −20 °C for 5 min or 3.7% paraformaldehyde, 5% acetic acid in PBS for 20 min (4). Since both of these methods also permeabilize membranes, skip to step 5.

[f] It is most convenient to use a vacuum suction device to remove the solution from the 6-well plate and to simultaneously add the new solution using a 10 ml pipette. It is crucial that the cells on the coverslip never dry out.

[g] We recommend Dumont GG forceps.

[h] Nuclei can be counterstained for easy visualization by use of DAPI or Hoechst 33342. After two washes as described in step 14, incubate coverslips for 5 min in 2 ml of 500 ng/ml DAPI or Hoechst 33342 and wash twice more as described in step 14. DAPI and Hoechst give a blue stain.

[i] Vectashield (Vector Laboratories) or mounting medium from Molecular Probes is recommended.

[j] A good indication that all excess mounting medium has been removed is if the coverslip can not be moved on the coverslide when gently pushed with the filter paper.

Protocol 1 is for adherent cells. Suspension cells can either be centrifuged onto a coverslip in a Cytospin centrifuge and then processed as described in Protocol 1. Alternatively, cells can be labelled in suspension according to Protocol 1. Start out with about $3 \times 10^6$ cells and at each step, spin the cells for 2 min at 3000 *g* and carefully remove the supernatant. Antibody incubations are done in a volume of 100 μl on a rocking platform. After the last wash resuspend the cells directly in 50 μl mounting medium and spot 15 μl on a microscopy slide and cover with an empty coverslip.

Protocol 1 can be used for detection of multiple proteins. Generally results are not compromised by the simultaneous incubation of multiple primary or secondary antibodies. If detection of multiple proteins is done stepwise it is recommended that steps 7–16 are carried out for the first antibody and then repeated for the second antibody. In either case it is essential to ensure that the primary antibodies used are from different species (polyclonal from rabbit and monoclonal from mouse are commonly used) and that the secondary antibodies do not cross-react between these species. When doing multiple labels the recommended sequence of fluorophores is fluorescein, Texas Red, Cy5, AMCA. Note that Cy5 emits in the infrared part of the spectrum and requires the use of an appropriate cooled CCD camera for detection. Indirect immunofluorescence can be combined with *in situ* hybridization methods to detect RNA or DNA (see Protocols 7–11).

## 3 GFP in fixed or living cells

Several autofluorescent proteins, most notably green fluorescent protein (GFP), from the jellyfish *A. victoria* allow detection of proteins in fixed or living cells without the need for a specific antibody (5). The fluorescent proteins are genetically encoded tags, which when fused to a cDNA of interest by standard recombinant DNA techniques, can be transiently or stably expressed and can be visualized directly in fixed or living cells. This method does not require the generation and characterization of antibodies, but only requires the subcloning of a cDNA in the appropriate position in a GFP expression vector. However, caution must be used when using GFPs for localization purposes. Overexpression of a fusion protein can dramatically alter both the function and localization of a protein. It is therefore desirable to only use fusion proteins for which correction function has been established. In general, for pure localization purposes it is recommended to always use indirect immunofluorescence (Protocol 1).

GFP fusion proteins can be observed in living cells. Because the microscopy on living cells is difficult, it is recommended that live cell imaging is only used when temporal or spatial redistributions of a protein need to be observed. When observing living cells, one needs to work fast. The exposure of cells to excitation light is damaging. Furthermore, since no antifade reagent is present, the fluorescence signal fades more quickly than in a fixed sample.

Visualization of GFP fusion proteins in fixed cells is compatible with detection of proteins by indirect immunofluorescence. After Protocol 2, step 4, perform a normal indirect immunofluorescence experiment (start at Protocol 1, step 5).

## Protocol 2

### Observation of GFP fusion proteins in fixed cells[a]

#### Equipment and reagents

- 22 × 22 mm coverslips
- Microscopy coverslides
- Filter paper
- PBS pH 7.4
- Paraformaldehyde (granular, Electron Microscopy Sciences)
- Mounting medium (Molecular Probes)

#### Method

1. Transfect cells with GFP fusion protein of interest and seed the cells on sterile coverslips.
2. Grow cells for desired length of time, typically 16–24 h.
3. Fix cells in 2% formaldehyde in PBS pH 7.4 for 15 min at room temperature.[b]
4. Quickly rinse cells in 4 ml PBS, and wash the coverslips in 2 ml PBS, twice for 5 min at room temperature.
5. Place 10 μl of mounting medium on a glass microscopy slide. Remove excess PBS by blotting the edges of the coverslip gently against a piece of filter paper.
6. Invert the coverslip cells facing down onto the mounting medium.
7. Remove excess mounting medium by holding two pieces of filter paper against two opposing edges of the coverslip until no more medium is drawn up by the filter paper.

[a] Protocol 2 applies to all versions of GFP, CFP (cyan), YFP (yellow), BFP (blue), and dsRed2 (red).
[b] Do not use methanol or acetic acid fixation as organic solvents destroy the autofluorescent properties of GFP.

## Protocol 3

### Observation of GFP fusion proteins in living cells

#### Equipment and reagents

- Nalgene Labtek II chambers (Nalgene) or glass-bottomed Petri dishes (Mattek)
- Phenol red-free cell growth medium (Gibco)

#### Method

1. Transfect cells with GFP fusion protein of interest and seed in Nalgene Labtek II chambers or in glass-bottomed chambers.
2. Grow cells for desired length of time, typically 16–24 h.
3. Replace growth medium with phenol red-free medium.[a]
4. Place chamber on inverted microscope and use appropriate filter sets to illuminate GFP.

[a] Phenol red gives significant background fluorescence.

## 4 Nuclear matrix preparation

The nuclear matrix is defined biochemically as a fraction of nucleoproteins resistant to sequential extraction with detergent, high salt, and DNase. Structurally, the nuclear matrix has been proposed to consist of an interconnected meshwork of filaments within the nucleus. Macromolecules involved in DNA replication, transcription, and RNA processing appear to be associated with the nuclear matrix. However, the physiological relevance of the nuclear matrix is unclear (6).

A nuclear matrix preparation reveals the extraction properties of a protein of interest. The nuclear matrix preparation is a three-step extraction procedure followed by conventional indirect immunofluorescence microscopy (see Protocol 1). Note that this protocol can also be applied to GFP fusion proteins, since the GFP fluorescence signal remains stable throughout the extraction procedure. The first fractionation step with detergent removes both the cytoplasmic and nuclear soluble proteins. The second step with high salt removes loosely bound proteins such as chromatin proteins (e.g. histone H1). The third step with DNase I extracts DNA. The resulting nuclear matrix includes most of the heterogeneous nuclear ribonucleoproteins (hnRNPs), hnRNAs, nuclear lamins, actin, and nuclear pore complex-associated proteins (7–10). The protocol given below can be used for adherent or suspension cells. An alternative protocol involving the encapsulation of cells within agarose microbeads is described elsewhere (11).

## Protocol 4

### *In situ* preparation of nuclear matrix for microscopy and immunoblotting

#### Equipment and reagents

- Tissue culture dishes (35 mm and 150 mm)
- Tissue culture plates (6-well)
- Glass coverslips (22 × 22 mm, No. 1.5) stored in 75% (v/v) ethanol
- Phosphate-buffered saline (PBS)
- 5 × SDS sample buffer
- 2% paraformaldehyde in PBS, prepare fresh and keep on ice
- Cytoskeleton (CSK) buffer: 100 mM NaCl, 300 mM sucrose, 10 mM Pipes pH 6.8, 3 mM $MgCl_2$, store in aliquots at −20 °C. Before use add the following: 0.5% Triton X-100, 1 mM PMSF (or AEBSF), 10 mM leupeptin, 10 μM pepstatin A, 15 μM E-64, 50 μM bestatin, 20 U/ml recombinant RNasin (Promega) ribonuclease inhibitor. After reconstitution keep on ice.
- DAPI (Molecular Probes)
- Extraction buffer: 250 mM ammonium sulfate, 300 mM sucrose, 10 mM Pipes pH 6.8, 3 mM $MgCl_2$, 0.5% Triton X-100, 1 mM PMSF (or AEBSF), 10 μM leupeptin, 10 μM pepstatin A, 15 μM E-64, 50 μM bestatin, 20 U/ml recombinant RNasin (Promega) ribonuclease inhibitor. Prepare fresh and keep on ice.
- Digestion buffer: 50 mM NaCl, 300 mM sucrose, 10 mM Pipes pH 6.8, 3 mM $MgCl_2$, store in aliquots at −20 °C. Before use add the following: 0.5% Triton X-100, 1 mM PMSF (or AEBSF), 10 μM leupeptin, 10 μM pepstatin A, 15 μM E-64, 50 μM bestatin, 20 U/ml recombinant RNasin (Promega) ribonuclease inhibitor, 200–500 U/ml RQ1 RNase-free[a] DNase (Promega). Prepare fresh and keep at room temperature.

**Protocol 4** continued

## Method

1. Remove coverslips from the 75% ethanol solution, flame to burn the ethanol off, and place them into 35 mm tissue culture dishes or into wells of a 6-well tissue culture plate.[b]
2. Plate the cells such that they reach 50–70% confluency at the day of experiment.[c]
3. Wash the cells twice with PBS at 4 °C.[d]
4. Gently add 3 ml of CSK buffer to each dish and incubate for 10 min at 4 °C.[e]
5. Remove the CSK buffer completely by tilting the dishes and aspirating the buffer with a glass capillary pipette.
6. Transfer coverslip No. 1 to a new well. Fix it by adding 3 ml of 2% paraformaldehyde and incubating for 15 min at room temperature.
7. Gently add 3 ml of extraction buffer to each dish and incubate for 5 min at 4 °C.
8. Remove the extraction buffer completely by tilting the dishes and aspirating the buffer with a glass capillary pipette.
9. Transfer coverslip No. 2 to a new well. Fix it by adding 3 ml of 2% paraformaldehyde and incubating for 15 min at room temperature.
10. Gently add 1 ml of digestion buffer and incubate for 30–60 min at room temperature.[f]
11. Remove the digestion buffer completely by tilting the dishes and aspirating the buffer with a glass capillary pipette.
12. Terminate the digestion by gently adding 3 ml of extraction buffer and incubate for 5 min at 4 °C.[g]
13. Remove the extraction buffer. Fix coverslips No. 3, 4, and 5 by adding 3 ml of 2% paraformaldehyde and incubating for 15 min at room temperature.
14. Process all samples for immunofluorescence microscopy (see Protocol 1).[h]

[a] It is essential that the DNase I is RNase-free.

[b] Prepare at least five coverslips per sample. The first coverslip will be extracted only with Triton X-100. The second coverslip will be extracted with Triton X-100 followed by ammonium sulfate. The third and fourth coverslips will be extracted with Triton X-100, ammonium sulfate, and digested with DNase I. The fourth coverslip is used as positive control (see below). The fifth coverslip is not extracted (control). It is generally convenient to use 6-well tissue culture plates when multiple samples need to be processed at the same time.

[c] This also applies to transiently transfected cells.

[d] If using suspension cells, approx. $10^6$ cells per extraction step will be adequate. Centrifuge the suspension cells at 650–1000 *g* for 5 min at 4 °C between steps 3–13. Resuspend the cells at the each extraction step in the same volume of buffer as the adherent cells.

[e] Incubate by simply placing the dishes on ice.

**Protocol 4** continued

[f] The duration of digestion and the necessary amount of DNase I will vary depending upon the cell type used and has to be optimized. A good starting point, particularly in mouse cell lines (heterochromatin forms dense structural domains) is to use 500 U/ml DNase I and three different incubation times (15, 30, and 60 min) at room temperature. Monitor the release of chromatin by staining the cells with DAPI for fluorescence microscopy (see Protocol 1). DNase I treatment must lead to a complete loss of DAPI stain. If the digestion is insufficient, use 30 and 60 min digestions both at 32 °C and 37 °C.

[g] For Western blot analysis, centrifuge the cells as above and measure the packed cell volume. In order to extract proteins from the nuclear matrix, add two to three times the packed cell volume of 5 × SDS sample buffer to a final concentration of 1 × SDS. Boil the samples for 5 min. Vortex or sonicate the samples (depending on the solubility of extract). Centrifuge the samples in a bench-top centrifuge at 16 000 g for 10 min at room temperature.

[h] It is very important to verify the integrity of nuclear matrix by processing coverslip No. 4 for indirect immunofluorescence microscopy. Use an antibody directed against one of the well-known matrix-associated proteins (e.g. NuMA). The hnRNPA1 and histone H1 proteins could be used as the negative controls for the detergent and high salt extraction steps respectively.

## 5 Visualization of transcription sites

Sites of RNA synthesis in the nucleus can be visualized by using the UTP nucleotide analogue BrUTP. All three RNA polymerases (RP) incorporate BrUTP into RNA. The incorporation efficiency is higher with RPII and RPI than RPIII. Short incorporation periods, if used as in the protocols described below will permit the detection of transcription initiation sites. Longer incorporation periods can be used to detect elongated transcripts. Although the rate of transcription in the presence of BrUTP is somewhat lower than in the presence of UTP, BrUTP incorporation does not induce any transcription artefacts (12–14).

The two protocols described here are designed to label and detect the transcription sites *in situ* or *in vivo*. In the first protocol, lightly permeabilized cells are allowed to incorporate BrUTP *in situ* for a short period of time. Cells are then fixed and transcription sites are detected by IF (Protocol 1). In the second protocol, BrUTP is microinjected into the cells *in vivo* and detected by IF. An alternative protocol involving the encapsulation of cells in agarose microbeads is described elsewhere (13). Transcription sites labelled with these protocols and visualized by microscopy show multiple distinct foci distributed throughout the nucleus and nucleolus. The number of sites detected will vary from few hundreds to few thousands depending on cell type. The intensity of the foci also differs from cell to cell in an asynchronous population of interphase cells.

When these protocols are used in conjunction with IF (Protocol 1) and *in situ* hybridization techniques (Protocols 9–11), one can simultaneously study the spatial relationships between transcription sites and a protein, a gene, or a nuclear domain of interest (12–14). It is also possible to determine whether the gene of interest is active or inactive. Both protocols can also be applied to GFP fusion proteins, since the GFP fluorescence signal remains stable throughout the transcription procedure.

## Protocol 5

# *In situ* labelling of newly-synthesized RNA with 5-bromouridine 5′-triphosphate and immunodetection of transcription sites by microscopy

### Equipment and reagents

- Tissue culture dishes (35 mm)
- Glass coverslips (22 × 22 mm, No. 1.5) stored in 75% (v/v) ethanol
- Phosphate-buffered saline (PBS)
- Trypan Blue dye 0.4% (Sigma)
- 2% paraformaldehyde in PBS, prepare fresh and keep on ice
- Permeabilization buffer: 20 mM Tris–HCl pH 7.4, 5 mM $MgCl_2$, 0.5 mM EGTA, 25% glycerol, 5–40 μg/ml digitonin (Calbiochem) or 0.02–0.1% Triton X-100 (Calbiochem), 1 mM PMSF, 20 U/ml recombinant RNasin (Promega) ribonuclease inhibitor. Prepare fresh, add the last two ingredients before use, and keep at room temperature.
- RNase A (Sigma)
- Transcription buffer: 100 mM KCl, 50 mM Tris–HCl pH 7.4, 10 mM $MgCl_2$, 0.5 mM EGTA, 25% glycerol, 2 mM ATP (Roche Molecular Biochemical), 0.5 mM CTP (Roche Molecular Biochemical), 0.5 mM GTP (Roche Molecular Biochemical), 0.5 mM BrUTP (Sigma), 1 mM PMSF, 20 U/ml recombinant RNasin (Promega) ribonuclease inhibitor. Prepare fresh, add the last two ingredients before use, and keep at room temperature.
- Monoclonal mouse antibodies to bromodeoxyuridine: clone BR-3 (Caltag), or clone BMC-9318 (Roche Molecular Biochemical), or clone Bu-33 (Sigma)
- Actinomycin D (Calbiochem) and α-amanitin (Sigma)
- DAPI nucleic acid stain (Molecular Probes)

### Method

1. Remove coverslips from the 75% ethanol solution, flame to burn the ethanol off, and place them into 35 mm tissue culture dishes.[a]
2. Plate the cells such that they reach 50–70% confluency at the day of experiment.[b]
3. Put the transcription buffer into the tissue culture incubator at 37 °C.[c]
4. Take the first dish out of tissue culture incubator and wash the cells twice with PBS at room temperature.[d]
5. Remove the PBS completely by tilting the dish and aspirating the PBS with a glass capillary pipette.
6. Gently add 2 ml of permeabilization buffer while lowering the tilted dish and incubate for 3 min at room temperature.[e]
7. Remove the permeabilization buffer completely by tilting the dish and aspirating it with a glass capillary pipette.
8. Gently add 1.5 ml of transcription buffer while lowering the tilted dish and incubate for 5 min at 37 °C.[f]
9. Remove the transcription buffer completely by tilting the dish and aspirating it with glass capillary pipette.

**Protocol 5** continued

**10** Gently add 1 ml of PBS while lowering the tilted dish.

**11** Remove PBS completely by tilting the dish and aspirating it with a glass capillary pipette.

**12** Fix the cells by adding 3 ml of 2% paraformaldehyde and incubating for 15 min at room temperature.

**13** Process the next sample (step 4), while cells are in the fixation solution.

**14** Remove the paraformaldehyde by aspirating it with a glass capillary pipette.

**15** Process all samples for IF (see Protocol 1).[g]

[a] It is generally more convenient to process only six dishes per experiment

[b] This also applies to transiently transfected cells.

[c] It is best to do the rest of the protocol on a laboratory bench very close to this tissue culture incubator.

[d] Process one dish at a time through steps 4–12.

[e] The cells may easily come off from coverslips so take great care between steps 4–14. The necessary concentration of detergent for a slight permeabilization will vary depending upon the cell type and has to be optimized in order to have a successful labelling. A good starting point is to monitor the level of permeabilization using Trypan Blue exclusion. Simply mix 100–200 μl of 0.4% Trypan Blue and 2 ml of permeabilization buffer containing 5–40 μg/ml digitonin or 0.02–0.1% Triton X-100 and add to the coverslips. Incubate for 3 min at room temperature and count the dark blue cells by light microscopy. Determine the detergent concentration that permeabilizes 50–75% of cells. Once the transcription sites are labelled successfully on a regular basis, it is very important to further optimize the detergent concentration until transcription site labelling is obtained with a minimum concentration of detergent.

[f] The duration of labelling may also vary depending upon the cell type and has to be optimized in order to have a successful labelling. As initial labelling times, we suggest using 5, 15, and 30 min both at 37 °C and room temperature. Once the transcription sites are labelled successfully on a regular basis, it is very important to further optimize the labelling period until transcription site labelling is achieved with minimum labelling time. A large and intense nucleolar staining is a good indication for the necessity of shorter labelling periods. The BrUTP concentration can also be lowered to 0.2 mM with identical results.

[g] Since digitonin does not permeabilize the nuclear membrane, if cells are being processed for indirect IF, they have to be permeabilized with Triton X-100 as in Protocol 1 in order for BrUTP antibody to gain access to the nucleus interior.

**Note**: The sensitivities of the commercially available monoclonal mouse antibodies to bromodeoxyuridine vary. The antibodies from Caltag and Roche Molecular Biochemical are far more sensitive than the antibody from Sigma under the conditions of these protocols. Good starting antibody dilutions for use in indirect IF are for the antibody from Caltag 1:200 in PBS, the antibody from Roche Molecular Biochemical 1:50 in PBS, and the antibody from Sigma 1:100 in PBS. These dilutions are optimized for use with the minimal cross-reactive secondary antibodies from Jackson ImmunoResearch Laboratories.

## Protocol 6

# *In vivo* labelling of newly-synthesized RNA with 5-bromouridine 5′-triphosphate and immunodetection of transcription sites by microscopy

### Equipment and reagents

- Tissue culture dishes (35 mm)
- Glass coverslips (22 × 22 mm, No. 1.5) stored in 75% (v/v) ethanol
- Phosphate-buffered saline (PBS)
- 2% paraformaldehyde in PBS, prepare fresh and keep on ice
- Actinomycin D and α-amanitin
- Microinjection buffer: 100 mM BrUTP (Sigma), 140 mM KCl, 2 mM Pipes pH 7.4; store in aliquots at −20 °C
- Monoclonal mouse antibodies to bromodeoxyuridine (see Protocol 5)
- RNase A
- DAPI nucleic acid stain

### Method

1. Remove coverslips from the 75% ethanol solution, flame to burn the ethanol off, and place them into 35 mm tissue culture dishes.
2. Plate the cells such that they reach 50–70% confluency at the day of experiment.[a]
3. Take the first dish out of the tissue culture incubator and place onto the stage of a fluorescence microscope equipped with a microinjection system.[b]
4. Inject the injection buffer into the cytoplasm of cells.[c]
5. Place the dish back in the tissue culture incubator and incubate for 5 min at 37 °C.[d]
6. Take the dish out of tissue culture incubator and gently add 1 ml of PBS while lowering the tilted dish.
7. Remove the PBS completely by tilting the dish and aspirating the PBS with a glass capillary pipette.
8. Fix the cells by adding 3 ml of 2% paraformaldehyde and incubating for 15 min at room temperature.
9. Remove the paraformaldehyde by aspirating it with a glass capillary pipette.
10. Process all of the samples for immunofluorescence microscopy (see Protocol 1).[e]

[a] This also applies to transiently transfected cells.

[b] Keep the cells in the growth medium. The growth medium change or PBS wash is not necessary at this stage.

[c] Inject always less than 5% of total cell volume. One indication that the cells are injected with more volume than necessary is the apparent change in cell morphology. Particularly, the edges of cells become uneven and cells lose their round shape. This can also be observed easily by staining the cells with DAPI after the paraformaldehyde fixation (step 8). To optimize the microinjection technique, replace the BrUTP with a fluorescent dye such as rhodamine in the injection buffer and inject different volumes of buffer into the cytoplasm, while observing the cells by fluorescence microscopy.

**Protocol 6** continued

[d] The duration of incubation may vary depending upon the cell type used and has to be optimized. A good starting point is to use four different incubation times (5, 15, 30, and 60 min). Once the transcription sites are labelled successfully on a regular basis, it is important to further optimize the labelling period until transcription site labelling is achieved with minimum labelling time.

[e] See footnote g in Protocol 5.

Several controls must be performed to verify whether the observed sites are genuine transcription sites:

(a) Do an experiment with UTP in the transcription and microinjection buffers instead of BrUTP. There should be no labelling detected by indirect immunofluorescence microscopy demonstrating the specificity of the antibody.

(b) Do a labelling with transcription and microinjection buffers containing 1–10 μg/ml α-amanitin or 5–10 μg/ml actinomycin D. If used in these concentrations, α-amanitin specifically inhibits the transcription by RNA polymerase (RNA pol) II. Thus, only the nucleolar transcription sites, but not the nuclear transcription sites should be detected. Actinomycin D, at the above concentration, inhibits transcription by all three RNA polymerases.

If no labelling is observed:

(a) Repeat the experiment by using either one of the other mouse monoclonal antibodies.

(b) Increase the duration of labelling.

(c) In Protocol 5, do the labelling at 37 °C.

(d) Increase the concentration of BrUTP to 1 mM.

(e) Optimize the concentration of detergent.

(f) Use a biotin–streptavidin fluorescent signal enhancement system (e.g. biotin-conjugated anti-mouse IgG and Texas Red dye-conjugated streptavidin) in indirect IF. These options can be used individually or in combination.

## 6 Fluorescence *in situ* hybridization

*In situ* hybridization (ISH) is a technique which allows visualization of specific nucleic acid sequences in cells through hybridization of labelled complementary sequence probe. ISH is used for diagnosis of infectious diseases, cytogenetic analysis, and cell biology studies of gene expression (4, 15, 16).

Probe sequences can be labelled with radioisotopes, enzymes, haptens such as biotin or digoxigenin, or fluorochromes. The most popular type of ISH is fluorescence *in situ* hybridization (FISH). FISH is usually performed with probes labelled with biotin or digoxigenin and then detected by a detection reagent conjugated to a fluorochrome. Probes can be labelled directly by incorporation of nucleotides conjugated with a fluorochrome, but the sensitivity of such probes is usually lower than indirect detection.

The length of the probe is critical for the hybridization efficiency and the background level. The probe has to be small enough to penetrate into the cell but can not be too large to prevent non-specific binding.

The *in situ* hybridization fixation procedure must preserve the morphological features of a cell and must make the targets accessible to the procedure, but not remove them. The fixation process must also not inhibit the diffusion of molecules such as avidin, haptens, antibodies, or enzymes, required to detect the labelled DNA or RNA. Precipitating fixatives, such as ethanol, ethanol/acetic acid (3:1), or methanol are very effective in preserving DNA. However, they result in poor morphology and they do not immobilize RNA very well, which may wash away during processing. Crosslinking reagents such as paraformaldehyde and glutaraldehyde preserve cellular morphology better, and firmly immobilize DNA and RNA in cells, but they tend to hinder the diffusion of probes. A compromise can be achieved by fixation in 3.7% paraformaldehyde/5% acetic acid (5).

Most commonly used detection probes are:

- nick translation probes (Protocol 7)
- antisense *in vitro* transcription probes (Protocol 8)
- oligonucleotide probes

FISH is compatible with detection of proteins by indirect immunofluorescence (16, 17).

## Protocol 7

## Probe generation by nick translation

### Reagents

- *E. coli* DNA polymerase I, nick translation grade (Roche Molecular Biochemicals)
- 1 mM biotin-16-dUTP (Roche Molecular Biochemicals)
- 100 mM dNTPs set (Roche Molecular Biochemicals)
- 2-mercaptoethanol (Sigma)
- Bovine serum albumin (BSA) (Sigma)
- Pre-chilled (−20 °C) absolute ethanol
- Nuclease-free water
- 10 × nick translation buffer: add 500 μl of 1 M Tris–HCl pH 8.0, 100 μl of 0.5 M $MgCl_2$, and 5 μl of 0.5 mg/ml BSA into a 1.5 ml test-tube; add 395 μl water to adjust the reaction volume to 1 ml
- Nucleotide stock: to a 1.5 ml test-tube add 10 μl of 10 mM dATP, 10 μl of 10 mM dCTP, 10 μl of 10 mM dGTP, 10 μl of 10 mM biotin-16-dUTP, and 4 μl of 1 M Tris–HCl pH 7.4; add 156 μl nuclease-free water to adjust the reaction volume to 200 μl
- 0.1 M 2-mercaptoethanol: to a 1.5 ml test-tube add 0.7 μl 2-mercaptoethanol; add 999.3 μl sterile water to adjust the final volume to 1 ml
- DNase I, RNase-free (10 U/μl) (Roche Molecular Biochemicals): make a stock solution of 10 mg/ml DNase I in 50% (v/v) glycerol which can be stored at −20 °C. Add 1 μl DNase I stock solution to 9 μl ice-cold water to make a stock solution (1 mg/ml). Dilute 1 μl DNase I (1 mg/ml) stock solution in 100 μl ice-cold water immediately before use; discard afterwards.

**Protocol 7** continued

### Method

1. Add the following to a 1.5 ml test-tube on ice:
   - 10 μl of 10 × nick translation buffer
   - 10 μl nucleotide stock
   - 10 μl 2-mercaptoethanol
   - 2 μg probe DNA
   - 1 μl DNase I (100 μg/ml)
   - 1 μl DNA polymerase I (10 U/μl)
   - 66 μl nuclease-free water

   Final reaction volume 100 μl.[a]
2. Centrifuge the test-tube briefly at 4 °C.
3. Immediately place the test-tube into a 16 °C water-bath. Incubate for 2 h.
4. Put the test-tube on ice immediately. Take 5 μl of sample from the test-tube, add agarose gel loading buffer, denature for 3 min at 95 °C, and run on a 1% agarose gel.[b]
5. To separate the probe from unincorporated nucleotides by ethanol precipitation add 1 μl of 0.1 M EDTA to stop the reaction, and then add 10 μl of 3 M sodium acetate pH 5.2 and 200 μl of pre-chilled (−20 °C) absolute ethanol.
6. Let the precipitate form for at least 20 min at −20 °C.
7. Centrifuge the tube at 13 000 g for 15 min at 4 °C. Very carefully discard the supernatant.
8. Dry the pellet under vacuum. Take care not to over dry the sample.
9. Resuspend the DNA in 20 μl nuclease-free water for a final concentration of ~100 ng/μl.
10. Store at −20 °C.

[a] The appropriate dilution of DNase I is critical in controlling the size of the labelled probe.

[b] The probe should run as a smear between 100–300 bp in size. If the probe is not the correct size, return the test-tube to the 16 °C water-bath for further digestion. If the probe is less than 100 nt in size use less DNase I or shorter incubation periods.

Probes can be generated by *in vitro* transcription reaction in the presence of labelled nucleotides. The target DNA fragment to be detected is cloned into the polylinker region of a transcription vector containing promoters for either SP6, T3, or T7 RNA polymerase. To synthesize an antisense 'run-off' transcript, a specific restriction endonuclease should be used to linearize the DNA template and to create a 5′ overhang on the antisense strand before transcription. The procedure described here incorporates one biotinylated nucleotide for approximately for every 20–25 nucleotides in the newly-synthesized transcript.

## Protocol 8

# Probe generation by *in vitro* transcription

### Reagents

- 10% SDS
- 10 mg/ml proteinase K (Roche Molecular Biochemicals)
- Phenol/chloroform/isoamyl alcohol (25:24:1) (Gibco)
- Ethanol
- SP6, or T3, or T7 polymerase, 20 U/ml (Roche Molecular Biochemicals)
- 1 mM biotin-16-UTP (Roche Molecular Biochemicals)
- 100 mM NTPs (Roche Molecular Biochemicals)

### Method

1. In a 1.5 ml test-tube perform a restriction digest on 20–25 μg of template DNA. Choose a specific restriction endonuclease which cuts the template ~200–300 nt downstream of the promoter region, to create an 5′ overhang to linearize the DNA template used for 'run-off' transcription.
2. After the restriction digest add 10 μl of 10% SDS, 2 μl of 10 mg/ml proteinase K, and add enough water to make a total reaction volume of 200 μl.
3. Incubate the tube for 30 min at 50 °C.
4. Add 8.5 μl of 0.5 M EDTA pH 8.0, 1.5 μl water, and 210 μl phenol/chloroform/isoamyl alcohol. Vortex for 1 min.
5. Centrifuge the tube at 13 000 g for 5 min at room temperature.
6. Transfer the upper aqueous layer into a new 1.5 ml test-tube. Add 200 μl phenol/chloroform/isoamyl alcohol. Vortex for 45 sec.
7. Centrifuge the tube as in step 5.
8. Transfer 180 μl of the upper aqueous layer to a clean tube. Add 20 μl of 3 M sodium acetate pH 5.2 and 500 μl pre-chilled (-20 °C) 100% ethanol. Let the precipitate form for at least 20 min at −20 °C.
9. Centrifuge the tube at 13 000 g for 30 min at 4 °C.
10. Discard the supernatant. Add 200 μl pre-chilled (−20 °C) 70% ethanol.
11. Centrifuge the tube as in step 9.
12. Discard the supernatant.
13. Dry the pellet under vacuum. Dissolve the pellet into 10 μl nuclease-free water. Adjust the final concentration to ~1 μg/μl. If desired, store at −20 °C.
14. Set up an *in vitro* transcription reaction by adding the following to a 1.5 ml microcentrifuge tube on ice:
    - 2 μl of 10 × transcription buffer (supplied with SP6, T3, or T7 RNA polymerase)
    - 2 μl of 10 mM ATP
    - 2 μl of 10 mM CTP
    - 2 μl of 10 mM GTP

**Protocol 8** continued

- 1.3 μl of 10 mM UTP
- 0.7 μl of 1 mM biotin-16-UTP
- 1 μg purified linearized DNA template
- 0.5 μl RNA pol (SP6, T3, or T7) (20 U/μl)
- Nuclease-free water to make a total reaction volume of 20 μl

**15** Mix the components and centrifuge briefly.

**16** Incubate the tube for 2 h at 37 °C.[a]

**17** Add 2 μl of 0.2 M EDTA pH 8.0 to the reaction tube to stop the reaction. Precipitate the DNA by adding 2 μl of 4 M LiCl and 60 μl pre-chilled (−20 °C) 100% ethanol. Let the precipitate form for at least 20 min at −20 °C.

**18** Centrifuge the tube at 13 000 g for 15 min at 4 °C.

**19** Discard the supernatant.

**20** Wash the pellet with 50 μl pre-chilled 70% (v/v) ethanol.

**21** Centrifuge the tube at 13 000 g for 10 min.

**22** Discard the supernatant.

**23** Dry the pellet under vacuum. Dissolve the RNA pellet for 30 min at 37 °C in 50 μl nuclease-free water plus 1 μl RNase inhibitor (20 U/μl).

**24** To estimate the yield of the transcript run an aliquot on an agarose gel beside an RNA standard of known concentration. Stain with ethidium bromide. Compare the relative intensity of staining between the labelled transcript and the known standard.[b]

[a] If you want to remove the template DNA, add 2 U DNase I, RNase-free to the test-tube, incubate for 15 min at 37 °C, and then go to step 4.

[b] If you are not going to use the labelled probe immediately, store the probe solution at −70 °C.

## Protocol 9

## Fluorescence *in situ* hybridization

### Equipment and reagents

- Glass coverslips, 22 × 22 mm
- Microscopy coverslide
- 6-well plates
- Forceps, Dumont GG (Electron Microscopy Sciences)
- Filter paper
- Parafilm
- Triton X-100 (Sigma)
- 4% paraformaldehyde (Electron Microscopy Sciences) in PBS pH 7.4
- Avidin–DCS–Texas Red or –fluorescein (Vector Laboratories)
- Mounting medium (Molecular Probes)
- Deionized formamide (Ambion)
- 10 mg/ml yeast tRNA (Sigma)
- Rubber cement

**Protocol 9** continued

## Method

1. Cells grown on a glass coverslips are fixed in 4% paraformaldehyde in PBS for 20 min at room temperature.[a]
2. Wash the cells with PBS three times for 5 min each at room temperature.
3. Permeabilize the cells with 0.2% Triton X-100 in PBS on ice for 5 min.
4. Wash the cells three times with PBS for 5 min each at room temperature.
5. Wash the cells with 2 × SSC for 5 min at room temperature.
6. Add the following to a 1.5 ml microcentrifuge tube at room temperature: 100 ng biotinylated nick translation probe or antisense riboprobe, 20 μg yeast tRNA.[b,c,d]
7. Dry it under vacuum.
8. Resuspend the pellet in 10 μl deionized formamide. Incubate the tube for 10 min at 75 °C. Put in water-ice slurry immediately for 5 min.
9. Add at room temperature 2 μl of 20 × SSC, 2 μl water, 2 μl of 50% dextran sulfate, 2 μl of 10 mM EDTA pH 8.0, 2 μl of 0.1 M Tris–HCl pH 7.2. Mix the components and centrifuge briefly.
10. Place 20 μl hybridization mixture onto each coverslip and seal with rubber cement.
11. Put the slide into a chamber moistened with 2 × SSC and incubate for 12–16 h at 42 °C.
12. After hybridization, remove coverslips and wash three times in 2 × SSC for 15 min each at 37 °C.
13. Wash once in 1 × SSC for 15 min at room temperature.
14. Incubate with avidin-conjugated fluorochrome (2 μg/ml) in dilution solution (4 × SSC/0.25% bovine serum albumin) for 1 h at room temperature.
15. Wash three times in 4 × SSC for 10 min each at room temperature.
16. Wash three times in PBS for 10 min each at room temperature.[e]
17. Mount coverslips in mounting medium.[f,g]

[a] Pre-extraction before fixation in CSK buffer with 0.5% Triton X-100 for 3 min on ice before fixation in paraformaldehyde (18) can increase the detection of target sequence.

[b] The following recipe is for a volume of 20 μl, which is sufficient for one hybridization reaction covering an area of 22 × 22 mm.

[c] If no signal is detected, the probe concentration can be increased up to 400 ng/20 μl.

[d] Hybridization without labelled probe should be performed as a control with each experiment.

[e] Optional. Wash briefly in water. Wash briefly in absolute ethanol and air dry.

[f] FISH is compatible with detection of proteins by immunofluorescence. After rinsing the cells in PBS, incubate the cells with primary antibody, then rinse in PBS, and incubate with appropriate secondary antibody (17).

[g] If background is a problem add 0.1% Triton X-100 to all wash solutions.

Labelled oligonucleotides (30–50-mers) can be used to detect RNAs. This method is particularly suited for the analysis of spliced versus unspliced RNA (16, 17). The efficiency of detection is generally lower than with nick translated probes.

## Protocol 10

## *In situ* hybridization with oligonucleotide probes

### Equipment and reagents

- See Protocol 9

### Method

1. Cells grown on glass coverslips are fixed in 4% paraformaldehyde in PBS for 20 min at room temperature.[a]
2. Wash the cells with PBS three times for 5 min each at room temperature.
3. Permeabilize the cells with 0.2% Triton X-100 in PBS on ice for 5 min.
4. Wash the cells with PBS three times for 5 min each at room temperature.
5. Wash the cells with 2 × SSC for 5 min at room temperature.
6. Add the following to a 1.5 ml microcentrifuge tube at room temperature: 4 μl of 20 × SSC, 4 μl of 50% dextran sulfate, ~1 μg/μl yeast tRNA, 0.2–1 pmol/μl labelled oligonucleotide probe. Enough nuclease-free water to make a total reaction volume of 20 μl.[a]
7. Place 20 μl hybridization mixture onto each coverslip and seal with rubber cement.[a]
8. Put the slide into a chamber moistened with 2 × SSC and incubate for at least 2–4 h at 37–42 °C.
9. After hybridization, remove coverslips and wash three times in 4 × SSC/0.1% Tween 20 for 5 min each at room temperature.
10. Block each coverslip in 250 μl of 4 × SSC/3% BSA/0.1% Tween 20 for 20 min at room temperature.
11. Incubate with avidin-conjugated fluorochrome (2 μg/ml) in 4 × SSC for 20 min at room temperature.
12. Wash three times in 4 × SSC/0.1% Tween 20 for 5 min each at 37 °C.
13. Mount the coverslip into mounting medium.[c]

[a] The following recipe is for a volume 20 μl, which is sufficient for one hybridization reaction covering an area of 22 × 22 mm.

[b] Hybridization without labelled probe should be performed as a control with each experiment.

[c] When *in situ* hybridization is followed by immunofluorescence, after rinsing the cells in PBS, incubate the cells with primary antibody, then rinse in PBS, and incubate with appropriate secondary antibody.

FISH can also used to detect specific DNA sequences or gene loci (4, 15). To make DNA accessible to the probe, the cellular DNA must be denatured by incubation in denaturation buffer.

## Protocol 11

## Fluorescence *in situ* hybridization to detect DNA sequences

### Equipment and reagents

- See Protocol 9

### Method

1. Fix subconfluent cells grown on glass coverslips in 3.7% paraformaldehyde/5% acetic acid in PBS for 15 min at room temperature.[a]
2. Wash the cells three times with PBS for 5 min each at room temperature.
3. Wash the cells twice with 2 × SSC for 5 min each at room temperature.
4. Denature coverslips in 70% formamide/2 × SSC for 7 min at 85 °C.[b]
5. Prepare the probe: denature 2 μl nick translated probe in 10 μl deionized formamide for 8 min at 95 °C.
6. Place immediately on ice.
7. Add hybridization buffer to give 50% formamide, 2 × SSC, 10% dextran sulfate, 1 mg/ml tRNA.
8. Place hybridization mixture onto each coverslip and seal with rubber cement.
9. Put the slide into a chamber moistened with 2 × SSC and incubate for 12–16 h at 37 °C.
10. After hybridization wash four times in 2 × SSC for 20 min each at room temperature.
11. Wash twice in 4 × SSC for 10 min each at room temperature.
12. Incubate with avidin-conjugated fluorochrome (2 μg/ml) in 4 × SSC for 1 h at room temperature.
13. Wash four times in 4 × SSC for 15 min each at room temperature.[c]
14. Mount the coverslip in mounting medium.

[a] Acetic acid permeabilizes membranes, therefore, no detergent permeabilization is required.

[b] Make sure the solution is at 85 °C when you add it to the coverslips. The most convenient way of doing this it to use a small water-bath and to float the coverslips in dishes.

[c] If background is a problem add 0.1% Triton X-100.

## 7 Chromosome preparation and staining

Chromosomes can not be seen with a light microscope in interphase cells, but during cell division they become condensed enough to be easily analysed under the microscope at ×1000 magnification. Chromosome studies are an important routine laboratory diagnostic procedure in certain patients with mental retardation, multiple birth defects, abnormal sexual development, infertility or multiple miscarriages, and in prenatal diagnosis. Cytogenetic analysis has become an important diagnostic, as well as prognostic tool in the evaluation of patients with cancer (19).

In this section we describe the protocols currently used in our laboratory for obtaining metaphase chromosome spreads, chromosome G-banding staining, and fluorescent chromosome painting.

Peripheral blood and bone marrow are the most common tissue types used to obtain chromosome preparations. Established cell culture lines are also suitable. Although specific techniques differ according to the type of tissue used, the general procedure for obtaining chromosome preparations involves:

(a) Tissue culture set-up and cell maintenance.

(b) Collecting cells with chromosomes in their condensed state by exposing to mitotic inhibitors, such as colchicine (Colcemid®) or nocodazole, which block formation of the spindle and arrest cell division at the metaphase stage.

(c) Harvest cells. This step involves exposing the cells to a hypotonic solution followed by fixation. It is important to perform this step very gently to obtain high quality preparations.

(d) Making the preparations by dropping the swollen cells onto a glass slide. This procedure breaks the cell membrane and allows the spreading of the chromosomes.

(e) Stain chromosome preparations for transmitted or fluorescent light microscopy.

Chromosomes in metaphase can be identified using several staining techniques. A variety of treatments involving denaturation and/or enzymatic digestion of the chromatin, followed by incorporation of dyes produce a chromosome-specific series of transverse alternating dark and light staining bands. This technique is called chromosome banding and is routinely used to identify each chromosome and characterize cytogenetic abnormalities using a transmitted light microscope. The most common staining technique is G-banding. G-bands are generated with Giemsa stain, but can also be produced with other dyes, such as Wright's stain. In G-bands, the dark bands tend to be heterochromatic, late replicating, and AT-rich. The light bands tend to be euchromatic, early replicating, and GC-rich. Alternatively R-banding or Q-banding techniques can be used. R-bands are approximately the reverse of G-bands and Q-bands correspond to fluorescent G-bands (20).

## Protocol 12

# Preparation of metaphase chromosome spreads from adherent cells or lymphocytes

### Equipment and reagents

- 75 $cm^2$ tissue culture flasks
- Centrifuge conical tubes
- Microscopy slides
- Pasteur pipettes
- Mouse interleukin-2 (Boehringer Mannheim)
- Phytohaemagglutinin M-form (Gibco BRL)
- KaryoMax Colcemid (Gibco BRL)
- Trypsin–EDTA (Gibco BRL)
- KCl solution: 0.056 M or 0.075 M in $H_2O$
- Methanol/glacial acetic acid (3:1) freshly made each time

### A. Method for adherent cells

1. Passage $10^6$ cells/ml to a 75 $cm^2$ flask containing 5 ml of regular growth medium two to three days prior to performing the chromosome spreads.
2. Feed cells with fresh medium 12–14 h prior to harvesting.
3. Add Colcemid to a final concentration of 0.02 μg/ml; incubate at 37 °C for 3–4 h.
4. Transfer the medium to a centrifuge tube and rinse the flask with 1.5 ml PBS (without $Ca^{2+}$ and $Mg^{2+}$). Detach the cells with 0.5 ml of trypsin. Collect the cells and rinse the flask twice with 1.5 ml PBS.
5. Spin down the cells at 160 *g* for 10 min.
6. Aspirate the supernatant. Leave about 0.5 ml of supernatant and resuspend the pellet tapping the bottom of the tube gently with your finger.
7. Add 5 ml of 0.056 M KCl solution. Pipette up and down to break any clumps. Leave at room temperature for 30 min.
8. Spin down the cells at 160 *g* for 10 min.
9. Aspirate the supernatant leaving 1 ml of the hypotonic solution. Resuspend pellet and add 1 ml of methanol/glacial acetic acid (3:1) solution.[a] Spin down the cells at 160 *g* for 10 min.
10. Aspirate the supernatant and resuspend the pellet in 3 ml fixative again. Add slowly and mix the cells as before for the first 1 ml.
11. Spin down the cells at 160 *g* for 10 min.
12. Aspirate the supernatant. Repeat steps 10 and 11 twice more.
13. Resuspend pellet in small volume of fixative (typically less than 500 μl), until the cell suspension looks slightly milky.
14. Take up a small quantity of cell suspension in the Pasteur pipette. Hold the pipette vertically with the end about 10 cm above the slide. Release a single drop from this height onto the slide. Let air dry.[b]

**Protocol 12** continued

### B. Method for suspension cells

1 Seed $10^6$ cells/ml in a 75 $cm^2$ flask containing 5 ml of regular growth medium supplemented with 100 μl of phytohaemagglutinin and/or 20 U/ml of mouse interleukin-2.[c]

2 Incubate the cell culture at 37 °C for 72 h. Shake the flask gently once or twice per day.

3 Add Colcemid (10 μg/ml) about 45 min before harvesting.

4 Transfer the cells into a centrifuge tube.

5 Continue as in part A, step 5, but use 0.075 M KCl solution.

[a] It is very important to disrupt the pellet before addition of the fixative, and to add the fixative only one or two drops at a time tapping the bottom of tube to mix the cells all the time.

[b] The room temperature and humidity are important factors to be considered when making the chromosome preparations. Chromosomes will spread poorly in a warm and dry environment. If necessary breath on the slide before dropping the suspension. Alternatively, place the dry slides in the freezer for 30–60 min and drop the cell suspension onto these cold and wet slides.

[c] Phytohaemagglutinin and mouse interleukin-2 are mitogenic agents. Human lymphocytes require only phytohaemagglutinin, whereas mouse lymphocytes grow better with a combination of both of them.

## Protocol 13

## G-banding using Wright's stain

### Equipment and reagents

- Coplin jars
- Staining slide rack
- Slide warmer
- 2 × SSC
- Sorensen's buffer pH 7.2: mix 49 ml of 0.06 M $Na_2HPO_4$ (8.52 g/litre) with 51 ml of 0.06 M $KH_2PO_4$ (8.16 g/litre)
- Mounting medium (Molecular Probes)
- Wright's stain (Sigma). To make a stain stock solution (2.5 mg/ml methanol) swirl 100 ml absolute methanol in a flask. Add 250 mg of powdered stain to swirling methanol. Continue to swirl for 30 min at moderate rate. Filter through Whatman No. 1 filter paper and aliquot into aluminium foil covered 10 ml bottles. Store away from heat and light.

### Method

1 Bake the chromosome preparations in an oven or slide warmer at 60 °C overnight.

2 Place the slides in a Coplin jar and fill it with 2 × SSC. Incubate the slides at 60 °C in a water-bath for 2–3 h. Discard the saline solution and rinse thoroughly at least 10 times with tap-water.

3 Let the slides air dry at room temperature for 12–16 h.

**Protocol 13** continued

4 Place the slides on a staining rack and cover the preparation with Sorensen's buffer for 45–50 sec.[a]

5 Discard the phosphate solution and cover the slide with Wright's stain solution for 60–90 sec (0.5 ml Wright's stain in 1.5 ml buffer).[b]

6 Rinse slides gently with tap-water,[c] let air dry, and mount if desired.

7 Observe under transmitted light microscope at ×1000 magnification.

[a] Staining time may need to be optimized. Always stain one slide first and check banding quality before staining additional slides.

[b] Adjust staining time if necessary, but it should be always less than 2 min. Otherwise it is recommended to prepare a new stain stock solution. Wright's stain will form a precipitate when added to buffer. Therefore, the two should not be mixed until just prior to flooding the slide.

[c] Be careful not to over-rinse slides since excessive rinsing will fade stain.

## 8 Fluorescent chromosome painting

Chromosome fluorescence *in situ* hybridization (FISH), also called chromosome painting, is a technique that combines cytogenetic methods for chromosome preparation and molecular genetic principles for DNA:DNA hybridization. Unlike other molecular DNA methods performed on free nucleic acids, FISH allows analysis of DNA *in situ*, that is, in its native, chromosomal form within the cell nucleus and in individual cells. Because individual chromosomes or chromosomal subregions can be painted differentially using specific probes, this technique allows metaphase and interphase chromosomes to be studied at a wider range of resolution levels than the conventional staining methods. Furthermore, microscopic evaluation of FISH experiments do not require trained cytogeneticists. Chromosome painting, on the other hand, has not only considerably improved the accuracy of clinical cytogenetic analyses, but has also been instrumental in studies of interphase nucleus organization and function (21, 22).

FISH uses nucleic acid probes. The probes are segments of fluorescently labelled DNA sequences designed to bind, or hybridize, with the target DNA of the specimen fixed to the glass slide. Three type of probes are commonly used:

(a) Locus-specific probes that hybridize to a particular gene or region of a chromosome (see Protocol 11).

(b) Centromeric and telomeric probes directed to the repetitive sequences found at the centromeres and telomeres of the chromosomes (see Protocol 11).

(c) Whole chromosome probes that identify a specific chromosome (Protocols 14–16). Chromosome paints are collections of smaller probes, each of which hybridizes to a different sequence along the length of a particular chromosome. Whole chromosome probes are particularly useful for examining chromosomal abnormalities, for example translocations.

The most common method to generate probes for chromosome painting is degenerate oligonucleotide-primed polymerase chain reaction (DOP-PCR). DOP-PCR is a method for non-specific, uniform amplification of DNA that uses degenerate oligonucleotides as primers. This design results in random priming on the target DNA. The DNA template in DOP-PCR reactions is DNA from flow sorted or microdissected chromosomes. For FISH experiments the DNA probe can be labelled by DOP-PCR by incorporation of nucleotides conjugated to haptens or fluorophores, such as biotin-dUTP or Spectrum Red-dUTP (23, 24). Alternatively the DNA probe labelling can be performed by nick translation (see Protocol 7). Chromosome painting probes are also commercially available (Vysis, Roche Molecular Biochemical), but are very expensive for routine use.

## Protocol 14

## Preparation of DNA probes for chromosome FISH: whole chromosome painting probes labelling by DOP-PCR

### Equipment and reagents

- PCR thermocycler
- Microcentrifuge
- Microcentrifuge tubes
- Source DNA (flow sorted or microdissected chromosomes)
- PCR buffer: 10 × Perkin Elmer without $MgCl_2$ (Roche Molecular Biochemical)
- 25 mM $MgCl_2$ solution Perkin Elmer (Roche Molecular Biochemical)
- 5 U/μl *Taq* DNA polymerase Perkin Elmer (Roche Molecular Biochemical)
- Deoxynucleotides: 100 mM dATP, dCTP, dGTP, and dTTP (Roche Molecular Biochemical)
- Stock dNTPs solution for labelling PCR contains 0.2 mM dATP, dCTP, dGTP, and 0.15 mM dTTP
- 0.06 mM fluorescein-dUTP (Boehringer Mannheim), or biotin-16-dUTP (Boehringer Mannheim), or Spectrum Red-dUTP (Vysis)
- Universal primer for human genomic DNA amplification: UN1 (Midland Certified Reagent Co.) Telenius [5′-CCGACTCGAGNNNNNNATGTGG-3′]
- Or universal primer for mouse genomic DNA amplification: 22-mer (Midland Certified Reagent Co.) [5′-CGG ACT CGA GNN NNN NTA CAC C-3′]
- Hi-Lo DNA marker (Minnesota Molecular)
- Tris acetate buffer
- 3 M sodium acetate pH 5.2

**Protocol 14** continued

## Method

1 Set up the following PCR reaction mix:

| Reagent | Quantity (μl) |
|---|---|
| • 10 × PCR buffer | 10 |
| • 25 mM $MgCl_2$ | 8 |
| • Stock dNTPs solution[a] | 5 |
| • Biotin-dUTP[a] | 5 |
| • $dH_2O$ | 65 |
| • DNA (100–150 ng/μl) | 4 |
| • 100 mM primer | 2 |
| • *Taq* polymerase (5 U/μl) | 1 |

2 Run the PCR reaction:

| Step | Temperature (°C) | Time (min) |
|---|---|---|
| 1 | 94 | 1 |
| 2 | 56 | 1 |
| 3 | 72 | 3<br>(+1 additional sec/cycle) |
| 4 | Steps 1–3 (29 times) | |
| 5 | 72 | 10 |

After the final step hold the PCR samples at 4 °C until they are used.

3 To analyse the DOP-PCR products, mix 8 μl of the reaction products with 2 μl agarose gel loading buffer.

4 Apply the sample to a 1% (w/v) agarose gel in 1 × TAE buffer.

5 Apply 10 μl of Hi-Lo DNA marker.

6 Run the gel for 45 min at 70 V/cm in 1 × TAE buffer.

7 Stain the gel with ethidium bromide and observe in UV transilluminator. In the reaction samples you should see a smear of DNA ranging from about 200–500 bp.

8 Add to the remnant of the DOP-PCR labelled DNA 0.1 vol. of 3 M sodium acetate pH 5.2 and 3 vol. of cold absolute ethanol.

9 Put tubes at −70 °C for 30 min.

10 Spin down the samples at 14 000 g in a microcentrifuge at 4 °C for 30 min.

11 Remove carefully the supernatant and dry the DNA under vacuum for 3 min.

12 Resuspend the DNA in sterile water at a final concentration of 100 ng/μl.

13 Store the labelled DNA at −20 °C.

[a] Note that the concentration of the dNTPs and labelled-dUTP varies.

The chromosome painting protocol follows this general outline:

(a) Preparation of the hybridization probe.

(b) *In situ* denaturation of the target DNA and hybridization.

(c) Post-hybridization washing and detection.

## Protocol 15

## Chromosome painting

### Equipment and reagents

- Slide warmer
- Water-bath or heat block
- 60 $mm^2$ cover glasses
- Coplin jars
- 1.5 ml microcentrifuge tubes
- Microcentrifuge
- Labelled probe (400–600 ng per hybridization)
- 1 mg/ml human Cot-1 DNA™ or 1 mg/ml mouse Cot-1 DNA (Gibco BRL)
- 9.7 mg/ml salmon testes DNA (Sigma)
- 3 M sodium acetate pH 5.2
- Formamide (molecular biology grade; Fisher Scientific)
- 70% (v/v) formamide in 2 × SSC pH 7.5
- 20% (w/v) dextran sulfate (Sigma) stock solution: dissolve 20 g dextran sulfate in 100 ml of 2 × SSC, adjust to pH 7.0, autoclave, store at −20 °C. Before using heat, vortex, and spin down at 14 000 *g* for a few seconds.
- 20 × SSC
- Absolute ethanol
- Blocking solution: 3% (w/v) bovine albumin (Sigma) in 4 × SSC/Tween 20
- Polyoxyethilene-sorbitan monolaurate (Tween 20) (Sigma)
- 1 mg/ml fluorescein–streptavidin (Amersham Pharmacia Biotech)
- DAPI (Sigma)
- Mounting medium (Molecular Probes)

### A. Preparation of the hybridization probe

1. Put in a 1.5 ml microcentrifuge tube 400–600 ng (≅ 4 μl) of probe DNA. Add 10 μg of the appropriate Cot-1 DNA (10 μl) and 10 μg (1 μl) of salmon testes DNA.
2. Add to the total volume 0.1 vol. of 3 M sodium acetate pH 5.2 and 3 vol. of ice-cold absolute ethanol.
3. Put the tubes at −70 °C for 30 min.
4. Spin down the samples at 14 000 *g* in a microcentrifuge at 4 °C for 30 min.
5. Remove carefully the supernatant and dry the DNA probe under vacuum for 2 min.
6. Resuspend the DNA probe in 5 μl deionized formamide pH 7.5 and incubate at 37 °C for 30 min. Vortex and spin down at 14 000 *g* for a few seconds every 10 min.
7. Add 5 μl of 20% (w/v) dextran sulfate in 2 × SSC.
8. Denature the DNA probe by heating at 80 °C for 5 min.
9. Spin down for a few seconds and incubate the DNA probe at 37 °C for 30 min to pre-anneal.
10. Keep on ice until used in part B, step 5.

**Protocol 15** continued

## B. *In situ* denaturation of the target DNA and hybridization

1. For chromosome slides denaturation place a 60 $mm^2$ cover glass on a slide warmer at 70 °C and apply 110 μl of 70% (v/v) formamide solution.[a]
2. Touch the denaturing solution with the preparation (sample side) and incubate on the hot plate for 1–2 min at 70 °C.[b]
3. Remove the cover glass and immerse immediately in ice-cold 70% (v/v) ethanol for 2 min to stop denaturation.
4. Dehydrate the slide through an ethanol series of 85% (v/v), and 100% (v/v) for 2 min each at room temperature, and then air dry.
5. Apply the previously denatured and pre-annealed probe. Cover with a 22 mm cover glass and seal the edges with rubber cement.
6. Incubate protected from light in a moist chamber at 37 °C overnight.

## C. Post-hybridization washing and detection

1. For post-hybridization washes and signal detection prepare Coplin jars containing the following solutions:
   (a) 50% (v/v) formamide in 2 × SSC pH 7.5.
   (b) 1 × SSC.
   (c) 4 × SSC/0.1% (v/v) Tween 20 (two jars).
   (d) 80 ng/ml DAPI staining solution in 2 × SCC (stock solution: 2 mg DAPI/10 ml sterile $H_2O$). Cover with aluminium foil.
   (e) Distilled $H_2O$.
2. Warm the washing solutions (a, b, and c, one jar) at 46 °C for 30–40 min.
3. Remove rubber cement and coverslip from the hybridized slide. Immerse the slide in jar (a). Agitate for a few seconds and leave it in the jar for 5 min.[c]
4. Transfer the slide to jar (b). Agitate for a few seconds and leave it in the jar for 5 min.
5. Transfer the slide to jar (c). Agitate for a few seconds and leave it in the jar for 5 min.
6. Dip the slide in 4 × SSC/Tween 20 at room temperature[d].
7. Apply 100 μl of blocking solution to a 60 $mm^2$ coverslip, touch the slides to coverslip. Incubate in moist chamber at 37 °C, protected from light, for 30 min.[e]
8. Rinse the blocking solution off by dipping the slide in 4 × SSC/Tween 20 for 3 min.
9. Apply to a 60 $mm^2$ coverslip 100 μl of fluorescein–streptavidin (dilution 1:100 in 4 × SSC/Tween 20/1% (w/v) bovine albumin). Touch the slide to coverslip. Incubate in moist chamber at 37 °C, protected from light, for 30 min.[f]
10. Rinse the slide in 4 × SSC/Tween 20 at room temperature, twice for 5 min, while shaking gently.

**Protocol 15** continued

11 Transfer the slide to jar (d). Counterstain with DAPI for 5 min at room temperature and protected from light.

12 Rinse the in $H_2O$, jar (e), for 5 min at room temperature.

13 Air dry and dehydrate the slide through an ethanol series of 70% (v/v), 85% (v/v), and 100% (v/v) for 2 min each at room temperature.

14 Air dry. Apply 30–35 μl antifade mounting medium and cover with 60 $mm^2$ coverslip. Store at 4 °C in the dark until analysis.

15 Visualize by eye using a fluorescence microscope

[a] Make the chromosome spreads as described in Protocol 1. Let them dry at room temperature for one to two days.

[b] Check the chromosome and nuclear morphology under the phase contrast microscope at low magnification, before and after the denaturation procedure, to optimize the time for each batch of chromosome preparations. Usually chromosomes acquire a fuzzy aspect due to over-denaturation. If so, reduce the time or the temperature by 3–5 °C. Otherwise increase the denaturation time. For interphase FISH the denaturation time usually is slightly longer. Sometimes it is necessary to remove remnants of cytoplasm (see Protocol 16).

[c] Do not let the slides dry during the washing and detection procedure.

[d] For probes directly labelled with a fluorescent nucleotide, i.e. Spectrum Red-dUTP, part C, steps 6–8 are avoided and the procedure continues at step 11.

[e] Spin down all the solution at 14 000 g for a few seconds before applying to the coverslips. In particular all fluorescent dyes.

[f] Multiple chromosomes can be detected simultaneously provided that distinct detection systems such as biotin/streptavidin and digoxigenin/anti-digoxigenin are used. Probes labelled with digoxigenin-UTP are detected with anti-digoxigenin–fluorescein (1:100 dilution).

Remnants of cytoplasm impair the access of the probe to the target DNA of metaphase and especially interphase chromosomes. To improve the hybridization quality a treatment with 70% acetic acid and/or mild pepsin treatment is frequently performed.

## Protocol 16

## Pre-treatment of chromosome or cells for FISH

### Equipment and reagents

- Coplin jars
- Slide warmer or water-bath
- Glacial acetic acid
- 1 N hydrochloric acid
- Pepsin (Sigma)
- 1 × PBS

**Protocol 16** continued

### A. Method for acetic acid treatment

1. Soak the slide in 70% (v/v) acetic acid for 40–60 sec.
2. Rinse the slide in PBS for 5 min at room temperature, shaking gently.
3. Dehydrate the slide through an ethanol series of 70% (v/v), 85% (v/v), and 100% (v/v) for 2 min each at room temperature, and then air dry.

### B. Method for pepsin treatment

1. Rinse the slide in PBS for 5 min.
2. Apply to a 60 $mm^2$ coverslip 100 μl of a 10 μg/ml solution of pepsin in 10 mM HCl. Touch the slide to the coverslip. Incubate in moist chamber at 37 °C for 5 min.
3. Rinse the slide in PBS for 5 min.
4. Dehydrate the slide trough an ethanol series of 70% (v/v), 85% (v/v), and 100% (v/v) for 2 min each at room temperature, and then air dry.

## References

1. Lamond, A. I. and Earnshaw, W. C. (1998). *Science*, **280**, 547.
2. Misteli, T. (2000). *J. Cell Sci.*, **113**, 1841.
3. Ried, T., Schrock, E., Ning, Y., and Wienberg, J. (1998). *Hum. Mol. Genet.*, **7**, 1619.
4. Dirks, R. W., Daniël, K. C., and Raap, A. K. (1995). *J. Cell Sci.*, **108**, 2565.
5. Tsien, R. Y. (1998). *Annu. Rev. Biochem.*, **67**, 509.
6. Pederson, T. (2000). *Mol. Biol. Cell*, **11**, 799.
7. Fey, E. G., Krochmalnic, G., and Penman, S. (1986). *J. Cell Biol.*, **102**, 1654.
8. Hozak, P., Sasseville, A. M., Raymond, Y., and Cook, P. R. (1995). *J. Cell Sci.*, **108**, 635.
9. Gonsior, S. M., Platz, S., Buchmeier, S., Scheer, U., Jockusch, B. M., and Hinssen, H. (1999). *J. Cell Sci.*, **112**, 797.
10. Javed, A., Guo, B., Hiebert, S., Choi, J. Y., Green, J., Zhao, S. C., *et al.* (2000). *J. Cell Sci.*, **113**, 2221.
11. Jackson, D. A. and Cook, P. R. (1988). *EMBO J.*, **7**, 3667.
12. Wansink, D. G., Schul, W., van der Kraan, I., van Steensel, B., van Driel, R., and de Jong, L. (1993). *J. Cell Biol.*, **122**, 283.
13. Jackson, D. A., Hassan, A. B., Errington, R. J., and Cook, P. R. (1993). *EMBO J.*, **12**, 1059.
14. Wei, X., Somanathan, S., Samarabandu, J., and Berezney, R. (1999). *J. Cell Biol.*, **146**, 543.
15. Xing, Y., Johnson, C. V., Moen, P. T., McNeil, J. A., and Lawrence, J. B. (1995). *J. Cell Biol.*, **131**, 1635.
16. Huang, S. and Spector, D. L. (1996). *J. Cell Biol.*, **131**, 719.
17. Misteli, T. and Spector, D. L. (1999). *Mol. Cell*, **3**, 697.
18. Carmo-Fonseca, M., Pepperkok, R., Carvalho, M. T., and Lamond, A. I. (1992). *J. Cell Biol.*, **117**, 1.
19. Vogel, F. and Motulsky, A. G. (1997). *Human genetics: problems and approach* (3rd edn). Springer–Verlag, Berlin.
20. Mitelman, F. (ed.) (1995). *ISCN An international system for human cytogenetic nomenclature*. S Karger, Basel.

21. Gray, J. W., Kallionemi, A., Kallionemi, O., Pallavicini, M., Waldman, F., and Pinkel, D. (1992). *Curr. Opin. Biotechnol.*, **3**, 623.
22. Cremer, C., Munkel, Ch., Granzow, M., Jauch, A., Dietzel, S., Eils, R., *et al.* (1996). *Mutation Res.*, **366**, 97.
23. Ried, T., Baldini, A., Rand, T. C., and Ward, D. C. (1992). *Proc. Natl. Acad. Sci. USA*, **89**, 1388.
24. Schröck, E., du Manoir, S., Veldman, T., Schoell, B., Wienberg, J., Ferguson-Smith, M. A., *et al.* (1996). *Science*, **273**, 494.

# Chapter 3
# Regulation of the cell cycle

Diana M. Gitig and Andrew Koff
Program in Molecular Biology and Cell Biology and Genetics, Cornell University Graduate School of Medical Sciences; and Memorial Sloan–Kettering Cancer Center, 1275 York Avenue, New York, NY 10021, USA.

## 1 Introduction

The smooth running of the mitotic cell cycle depends on the activity of a family of serine–threonine kinases, the cyclin-dependent kinases (cdks). Agents that regulate the cell cycle ultimately impinge upon the activity of these proteins, either directly (by promoting their degradation or inducing the expression of an inhibitor) or indirectly (by arresting the cell cycle at a point before a particular cdk is active). The cdks are under very strict regulation, as might be expected for enzymes responsible for such a fundamental cellular process. These kinases are activated and inactivated in a sequential manner over the course of the mitotic cycle and their phosphorylation of various substrates allows cells to progress through the mitotic cycle (1). To date, the most extensively characterized of these substrates is the retinoblastoma protein, pRb, whose phosphorylation first by cyclin D/cdk4,6 in early G1 phase and then by cyclin E/cdk2 later in G1 precedes the commencement of DNA synthesis (2, 3) (Figure 1). Other substrates that have been identified are listed in Table 1.

Many mechanisms that either activate or inhibit the cdks, and therefore either promote or arrest progression through the cell cycle, have been defined over the past decade. However, since the signal transduction pathways emanating from extracellular mitogens and antimitogens are complicated, there may yet be novel mechanisms of cell cycle regulation remaining to be elucidated. In this chapter we will outline three different approaches used to study the cell cycle and its regulation and the systems which exemplify them: the cell biology approach, using the tissue culture system; genetics, via knockout mice; and biochemistry, both *in vitro* using purified recombinant proteins and *in vivo* with cellular extracts.

## 2 Cell biology: detection of cell cycle arrests

As is often the case with important biological processes, cell cycle events and the molecules that control them tend to be discovered only when things go awry.

Thus, a frequent indicator that cell cycle regulators are being acted upon is cell cycle arrest or the continuation of growth when conditions dictate that the cell cycle should be arrested. The functions of two kip family members were identified in this manner; $p27^{Kip1}$ was isolated from MvLu1 cells treated with TGF-β, which were arrested in G1 despite the presence of cyclin E/cdk2 com-

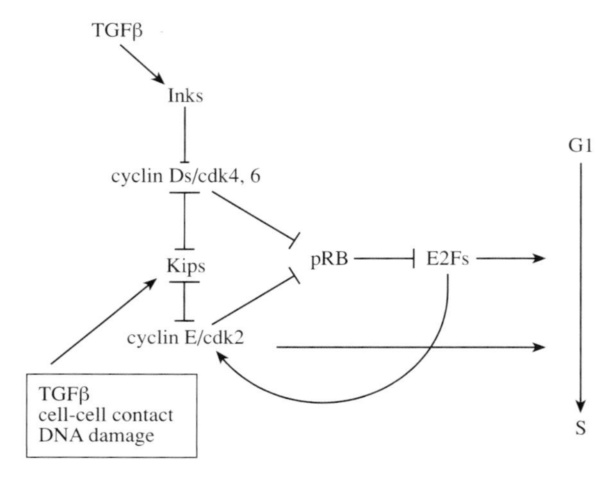


**Figure 1** The Rb pathway.

**Table 1** Substrates of cyclin/cdk complexes

| Protein | Activity | References |
|---|---|---|
| pRb, p107, p130 | Phosphorylation of these tumour suppressors by cyclin D/cdk4,6 and cyclin E/cdk2 during G1 phase causes them to dissociate from E2Fs and HDACS, thereby allowing E2F responsive genes to be expressed. | 2–4 |
| E2F(1–6), DP(1, 2) | Phosphorylation further activates these transcription factors, which up-regulate the expression of many genes necessary for S phase, including cyclin E and cdc6. | 5 |
| cdc6 | cdc6 is needed for formation of the preRC and, as such, is essential for initiation of DNA replication. Phosphorylation by cyclin A/cdk2 causes it to translocate from the nucleus to the cytoplasm, perhaps to prevent re-replication. | 6 |
| NPAT | Unknown. | 7 |
| RFA, human SSB | One subunit of this trimeric single-stranded binding protein is phosphorylated by cdks. | 5 |
| Histone H1 | This convenient *in vitro* substrate for cdk2 was one of the first identified. | 5 |
| $p27^{Kip1}$ | Although an inhibitor of cyclin E/cdk2, p27 is also one of its substrates. Phosphorylation prepares p27 for ubiquitination and proteolysis. | 8 |
| Lamins | Phosphorylation by the B-type cyclins correlates with dissolution of the laminar structure at mitosis. | 5 |

plexes at the same levels observed in untreated cells (9, 10); and p21$^{Cip1}$'s function as a regulator of the G1 DNA damage checkpoint was confirmed by the finding that p21-/- MEFs have a compromised ability to arrest in G1 in response to ionizing radiation and other DNA damaging agents (12, 13). It is often therefore of utmost importance to determine the cell cycle profile of a population of cells under certain conditions (see Protocol 1).

## Protocol 1

## Propidium iodide staining[a]

### Reagents

- PI stain: 10 mg propidium iodide, 100 μl Triton X-100, 3.7 mg EDTA (Sigma, Cat. No. E5134). Add phosphate-buffered saline (PBS) to 100 ml.
- RNase solution (200 U/ml): 10 mg RNase A, 5 ml PBS. Combine in a 15 ml tube, heat to 75 °C for 20 min, and cool to room temperature. Store in 1 ml aliquots at −20 °C.

### Method

1. Chill PBS, methanol, and samples on ice.
2. Prepare single cell suspensions of samples in ice-cold PBS.
3. Gently mix each suspension and count in a haemocytometer or Coulter counter. Adjust the concentration to 1–2 $\times$ $10^6$ cells/ml with ice-cold PBS.
4. Transfer 1 ml of each sample to a labelled 12 $\times$ 75 mm test-tube.
5. To fix the cells, add cold methanol dropwise from a Pasteur pipette while mixing the cells on a vortex on setting 4. Add 2 ml of methanol to 1 ml of the cell suspension.
6. Incubate the tubes on ice for at least 30 min. Samples will remain stable for up to one week at 4 °C in this methanol solution.
7. Centrifuge at 300 g for 5 min. Aspirate the supernatant with a Pasteur pipette.
8. Add 500 μl of PI stain solution to each tube and vortex gently to resuspend. Add 500 μl of RNase solution (final concentration = 100 U/ml) and vortex to mix.
9. Incubate at room temperature for 30 min in the dark.
10. Filter samples through 35 μm nylon mesh.
11. Keep the samples at 4 °C in the dark until you analyse them; they will keep overnight. Propidium iodide fluoresces primarily at wavelengths above 610 nm.

[a] Adapted from ref. 11.

Propidium iodide stains nucleic acids, and thus can be used (following RNase treatment) to determine the proportion of cells of a given population in each phase of the cell cycle. It can be used to stain either whole cells or nuclei. This technique is often used to demonstrate a block in a particular phase of the cell cycle following specific treatments. G1 cells have a 2n DNA content, G2/M cells have a 4n DNA content, and S phase cells have an intermediate amount. Apoptotic cells, which degrade their DNA, have a DNA content less than 2n; and cells in which M has been uncoupled from S and cells exhibiting genomic instability for other reasons may have DNA contents exceeding 4n.

In contrast to propidium iodide, which stains all of the nucleic acids in a cell, tritiated thymidine is incorporated only into DNA that is being replicated. Thus, thymidine incorporation is often used in combination with PI staining to ascertain what percentage of a cell population is in S phase or to confirm that cells are indeed growth arrested. As described in Protocol 2, thymidine incorporation can be measured over a time-course, which allows the rate of DNA synthesis to be determined. This is an important benefit of this technique, because it may reveal that despite identical cell cycle profiles as determined by PI staining, certain cells (treated, mutant, etc.) are now traversing S phase faster or slower than controls.

## Protocol 2

## [$^3$H]Tdr incorporation[a]

### Reagents

- 0.5% [$^3$H]Tdr labelled medium
- 10% TCA
- 10 × trypsin stock solution (Gibco BRL, Cat. No. 15400-054)

### Method

1. Seed 1.5 × $10^5$ cells/well in 3 ml in a 6-well TC plate and let them grow overnight. 3-5 × $10^5$ cells is the optimal number of cells for [$^3$H]Tdr labelling; the number of cells seeded per well should be adjusted according to cell doubling time.
2. Prepare 0.5% [$^3$H]Tdr labelled medium.
3. Aspirate the medium and add 1 ml labelled medium to each well or dish. Incubate at 37 °C for 2 h.
4. Aspirate the medium and wash three times with PBS.
5. Add 700 μl of 1 × trypsin to each well. After harvesting the trypsinized cells count them so the amount of [$^3$H]Tdr incorporated/cell can be calculated later.
6. Add 800 μl of cold 10% TCA to the harvested cells; vortex the tubes, and let them sit on ice for 30 min.
7. Spin at 4 °C for 15 min.
8. Wash with 1 ml cold 10% TCA and spin at 4 °C for 15 min.
9. Resuspend the pellet in 200 μl of 2 M NaOH.
10. Incubate in a 65 °C water-bath for 10-20 min.

**Protocol 2** continued

11. Add the sample to a scintillation vial containing 3 ml scintillation fluid; vortex well, and incubate at room temperature for 30 min.
12. Count in a scintillation counter. Cells entering S phase from G1 should give results of thousands of cpm.

[a] This protocol has been optimized for exponentially growing mouse embryonic fibroblasts (MEFs).

A caveat of [$^3$H]Tdr incorporation is that it cannot distinguish between a change in the *rate* of cells traversing S phase and a change in the *number* of cells traversing S phase. Thus, it is often combined with either PI staining and FACS or with BrdUrd (bromo-deoxyuridine, a thymidine analogue) staining. Like tritiated thymidine, BrdUrd is incorporated into DNA as it is being replicated. However, it can be used to label individual cells, rather than an entire population of cells, and thus it can be used to calculate the percentage of cells in a given population that are traversing S phase. BrdUrd labelling is usually used to determine the rate at which cells *change* from one proliferative state to another (i.e. the length of time it takes cells to move from G1 to S phase, or G0 to G1); however, it can also be used to determine the rate at which cells *traverse* a particular phase (i.e. the length of time cells spend in G0 or G1). Follow manufacturer's instructions for BrdUrd labelling; we have achieved good results with Boehringer BrdUrd. For cytokinetic studies, please refer to refs 14–18.

## 3 Genetics: knockout and transgenic mice

The information gleaned from knockout and transgenic mice has been invaluable in shaping our perceptions of how cell cycle regulatory proteins function during the development of an organism. Many surprising phenotypes have been obtained which have greatly enhanced our growing understanding of the cell cycle. The following examples are illustrative: since p21$^{Cip1}$ is a cdk inhibitor that is induced by p53 and arrests cells in G1 phase it was thought to be a tumour suppressor, but p21 null mice do not exhibit increased tumour incidence or shortened life span (12, 13); cyclin D-associated kinase activity was considered necessary for entry into S phase (19), but when cyclin E was knocked-in to the cyclin D1 locus, replacing all of cyclin D1's associated kinase activity, there were no ramifications in terms of cell doubling times or cell cycle profiles (20); and although loss of heterozygosity (LOH) at the Rb locus in humans yields retinoblastoma (hence the name of the protein), this same LOH event in mice yields pituitary tumours (21–23). The mice that are being generated will undoubtedly continue to raise as many questions as they answer. Since protocols for generating knockout and transgenic mice are beyond the scope of this chapter, we instead present a summary of the mice created to date and what we have learned from each (Table 2).

**Table 2** Knockout and transgenic mice of cell cycle-related proteins—cyclins, cdks, their inhibitors, and their substrates

| Mutation | Phenotype | References |
|---|---|---|
| pRb–/– | Lethality between embryonic day 13.5–15.5. Defects in neurogenesis and haematopoiesis. | 21–23 |
| pRb+/– | Animals are viable but develop thyroid and pituitary tumours. Life span is reduced to approximately 11 months. | 24, 25 |
| p130–/– | Viable in 129C57 mixed background. Non-viable in Balb/cJ. | 26, 27 |
| p107–/– | Viable in 129C57 mixed background. 50% smaller mice and ectopic myeloid hyperplasia in the spleen and liver in Balb/cJ. | 28, 29 |
| p130–/–, p107+/– | Same as p130–/–. | 26 |
| p130+/–, p107–/– | Smaller. | 26 |
| p130–/–, p107–/– | Neonatal lethality and impaired chondrocyte differentiation. | 26 |
| pRb–/–, p107–/– | Earlier lethality (e11.5) than pRb–/–. | 29 |
| pRb+/–, p107–/– | Growth retardation, increased mortality, and retinal dysplasia. | 29 |
| p21–/– | No evident phenotype, but MEFs have a defective G1 checkpoint. | 12, 13 |
| p27–/– | Gigantism, organomegalia, female infertility, pituitary hyperplasia, and deafness. | 31–33 |
| p57–/– | Some embryonic and neonatal lethality. Altered cell proliferation and apoptosis in several tissues. | 34, 35 |
| p21–/–, p57–/– | Altered lung development, enhanced skeletal abnormalities, and skeletal muscle differentiation failure. | 35 |
| p27–/–, p57–/– | Increased embryonic lethality, placental defects, and more severe lens alterations than individual nulls. | 36 |
| p16–/– | Increased tumour predisposition. | 37 |
| p15–/– | Viable. | 38 |
| p18–/– | Gigantism, organomegalia, and pituitary hyperplasia. | 39 |
| p19–/– | Testicular atrophy. | 40 |
| p15–/–, p18–/– | Infertility. | 38, 41 |
| p18–/–, p27–/– | Increased organomegalia and earlier onset of pituitary adenomas than individual nulls. | 39 |
| p19–/–, p27–/– | Postnatal lethality at day 18. Ectopic neuronal divisions and apoptosis leading to neurological defects. | 42 |
| p21–/–, pRb+/– | Decreased survival and earlier appearance of pituitary adenocarcinomas than in individual mutants. | 13 |
| p27–/–, pRb+/– | Decreased survival and pituitary tumours arise earlier and are more aggressive than in individual mutants. | 43 |
| p27–/–, p130–/– | Viable and fertile. | 44 |
| Cyclin D1–/– | Reduced body size, reduced viability, reduced retinal cell number, and neurological impairment. | 45 |
| Cyclin D2–/– | Female infertility, decreased granule cell number, and absence of stellate interneurons. | 46 |

**Table 2** (Continued)

| Mutation | Phenotype | References |
|---|---|---|
| cdk4–/– | Growth retardation, reproductive dysfunction, and diabetes mellitus. | 47, 48 |
| Cyclin E knock-in to cyclin D1 locus | Complete rescue of cyclin D1–/– phenotypes. | 20 |
| cdk4R24C | When this constitutively active cdk4 (it cannot be inhibited by $p16^{Ink4a}$) is substituted for wild-type, Rb is hyperphosphorylated, mice are 5–10% larger than wild-types, and pancreatic islet hyperplasia is observed. | 47 |

# 4 Biochemistry

## 4.1 *In vitro*: recombinant proteins

Because many cell cycle proteins are expressed only transiently during the cell cycle, and because the complicated cellular context often renders it difficult to observe events of particular interest, it can be useful to study the interactions between these molecules in an *in vitro* system using partially purified components. Cyclin/cdk complexes are most often produced in baculovirus-infected cells, as bacteria are unable to support the post-translational assembly and activating phosphorylations they require. The D-type cyclins in particular seem to need other factors to assemble with their catalytic subunits, cdk4 and cdk6 (49, 50). Thus, rather than individually infecting cells with the cyclin virus and the cdk virus and then mixing the lysates, cells are co-infected with the cyclin virus and the cdk virus. Kip class cdk inhibitors and a fragment of the cdk substrate Rb, however, can be purified from *E. coli* and mixed with the baculolysates to examine the interactions between the proteins. A method for preparing cyclin/cdk complexes in baculolysates is described in Protocol 3.

### Protocol 3

### Preparation of cyclin/cdk complexes in baculolysates[a]

#### Reagents

- Recombinant baculoviruses
- Grace's complete insect medium
- Cdk2 buffer: 10 mM Hepes–KOH pH 7.4, 10 mM NaCl, 1 mM EDTA, 1 mM phenylmethylsulfonyl fluoride (PMSF), 1 μg/ml leupeptin
- Cyclin D buffer: 50 mM Hepes–KOH pH 7.5, 10 mM $MgCl_2$, 1 mM dithiothreitol (DTT), 2.5 mM EGTA, 0.1 mM PMSF, 5 μg/ml aprotinin, 10 mM β-glycerophosphate, 0.1 mM $Na_3VO_4$, 0.1 mM NaF

**Protocol 3** continued

### A. Baculoviral infection

1. Seed $1.5 \times 10^7$ Sf9, Sf21, or Hi5 cells in 100 mm plates. Allow to attach for 30 min at room temperature.
2. Remove the medium and add the appropriate combination of cyclin and cdk baculovirus at a multiplicity of infection of 5–20 pfu/cell. Let the plates sit at room temperature for 1 h, swirling every 10 min.
3. Add Grace's complete medium to a total volume of 10 ml. Incubate at 27 °C for 48–72 h.

### B. Preparation of baculolysates

1. Collect the cells in a containment hood, either blowing them off with a pipette or scraping them off with a rubber spatula.
2. Spin the cells down at 700 r.p.m. at 4 °C for 5 min in a Sorvall RT6000D rotor. Store the virus-containing supernatant at 4 °C, protected from light. Wash the pellet in 10 ml of ice-cold PBS.
3. Resuspend the pellet in 400 μl/plate of the appropriate lysis buffer (see Reagents).
4. Dounce homogenize, using a tight pestle and the following pattern: seven strokes, 2 min rest on ice; three strokes, 2 min rest on ice; three strokes.
5. Add NaCl to a final concentration of 150 mM.
6. Microcentrifuge at 4 °C for 20 min. Transfer the supernatant to a new tube.
7. Measure the protein concentration by the Bradford assay. Ensure that you have isolated active cyclin/cdk complexes by doing the appropriate kinase assay (see Protocols 5 and 6) and store the lysates at −80 °C.

[a] Adapted from refs 51 and 52.

## 4.2 *In vivo*: extract reconstitution assays

Many important cell cycle proteins and their functions were discovered using extracts made from differently treated cells that were either mixed together or complemented with added proteins. For example, $p27^{Kip1}$ was identified using this technique (10). Extract reconstitution assays, such as those described in Protocol 4, can be used to measure the ability of a particular extract (representing a certain cell type, phase of the cell cycle, drug treated cell, or fraction of an extract) to activate, inhibit, degrade, assemble, or stabilize an exogenously added protein or proteins. Once the activity is identified in an extract, it can then be isolated, purified, and cloned. Conversely, the exogenously added protein may 'rescue' an activity impaired in the extract by permitting its occurrence, thereby determining what is blocking that event in the extract. The lysis method and buffer used to make cell extracts depend both on the type of cells and the purpose of the extracts. Cytoplasmic extracts prepared by Dounce homogenization may be used, but we have had the most success lysing whole cells in a hypotonic lysis buffer. The resulting lysates can be stored at −80 °C for years and subsequently used for immunoblots, immunoprecipitations, histone H1 and Rb kinase assays, and reconstitution experiments.

## Protocol 4

## Extract reconstitution assays

### Reagents

- HKM: 20 mM Hepes–KOH pH 7.5, 5 mM KCl, 0.5 mM $MgCl_2$. Autoclave and store in 10 ml aliquots at −20 °C.
- ATP regenerating system: 3 mM ATP from a 0.3 M stock in $H_2O$, 0.04 mg creatine phosphokinase from a 2 mg/ml stock in 50% glycerol, 0.04 M phosphocreatine from a 0.4 M stock in $H_2O$. All working stocks should be stored at −20 °C; long-term storage is at −80 °C.
- NP40-RIPA: 50 mM Tris–HCl pH 7.5, 250 mM NaCl, 0.5% NP-40, 5 mM EDTA pH 8.0, 1 mM PMSF,[a] 50 mM NaF,[a] 3 mM $Na_3VO_4$,[a] 10 μg/ml soybean trypsin inhibitor,[a] 10 μg/ml aprotinin,[a] 10 mM β-glycerophosphate[a]
- Tween-LB: 50 mM Hepes–KOH pH 7.5, 150 mM NaCl, 1 mM EDTA pH 8.0, 2.5 mM EGTA, 1 mM DTT,[a] 0.1% Tween-20, 10% glycerol, 10 mM β-glycerophosphate,[a] 1 mM NaF,[a] 0.1 mM $Na_3VO_4$,[a] 0.2 mM PMSF,[a] 10 μg/ml aprotinin,[a] 10 μg/ml leupeptin[a]

### A. Preparation of HKM extracts

1. If cells are adherent, trypsinize them. Collect the cells in ice-cold complete medium (5 ml per 150 mm plate).
2. Collect cells by centrifugation at 1000 g at 4 °C for 5 min.
3. Resuspend the cell pellet in 10 ml of cold PBS and transfer to a 15 ml centrifuge tube.
4. Collect cells by centrifugation at 1000 g at 4 °C for 5 min. Wash twice more with PBS.
5. Resuspend the cells in 1 ml of cold PBS and transfer them to an Eppendorf tube with a 1 ml Pipetman. Cut off the end of the tip to make a wider hole so the cells are not sheared.
6. Collect the cells in a microcentrifuge at 4 °C, but do not let the microcentrifuge achieve full speed as this will lyse the cells.
7. Aspirate the supernatant and approximate the volume of the cell pellet by comparing it by eye to a known volume in an equivalent tube. Be conservative; it is preferable to underestimate the volume of lysis buffer and end up with more concentrated lysates. Ideally, lysates should have a concentration of between 10–25 mg/ml.
8. Add 1.2 vol. of HKM buffer.

**Protocol 4** continued

9 Adjust to 2 mM PMSF (based on the volume of HKM buffer) from a 0.1 M stock prepared in ethanol.

10 Adjust to 0.5 mM DTT (based on the volume of HKM buffer) from a 0.2 M stock stored at −20 °C.

11 If the cells were trypsinized, add 1 μg/ml of soybean trypsin inhibitor from a 10 mg/ml stock.

12 Vortex quickly to resuspend the cell pellet, wrapping the top of the tube in Parafilm to avoid spills.

13 Sonicate at 20–30% power using a cuphorn sonicator/ 550 Sonic Dismembrator (Fisher) filled with iced water using the following guidelines:

- 1–200 μl total volume 2 min
- 2–300 μl 2.5 min
- 3–400 μl 3 min
- 4–700 μl 4 min
- >700 μl divide in aliquots

Sonicate for 1 min bursts, cooling on ice for 30 sec between pulses.

14 Pellet the cellular debris at 13 000 g at 4 °C for 10 min. After complete sonication the pellet should be about half the original size.

15 Transfer the supernatant to an ultracentrifuge tube, measure its volume, and adjust to 0.1 M NaCl.

16 Clarify the extract by ultracentrifugation at 100 000 g at 4 °C for 10 min.

17 Determine the protein concentration by the Bradford assay. Aliquot the supernatant and store at −80 °C.

## B. Reconstitution assays

1 To 50 μg of extract add the desired combination of cyclins, cdks, inhibitors, and/or other proteins in the form of Sf9 cell lysates, recombinant proteins from bacteria, rabbit reticulocyte lysates, or protein purified from cell extracts. Alternatively, different extracts (i.e. in different phases of the cell cycle or differentially drug treated) can be mixed in different ratios to determine which is dominant.

2 Incubate at 37 °C for 30 min with ATP and an ATP regenerating system.

3 Adjust conditions to NP40-RIPA or Tween-LB, as required.

4 Perform immunoprecipitation, histone H1 kinase assay, and/or Western blot, as desired. If performing an Rb kinase assay, adjust conditions to Tween-LB rather than NP40-RIPA.

[a] These protease inhibitors should be added immediately before use.

Progression through the cell cycle is absolutely dependent on the catalytic activity of cyclin/cdk complexes. Antimitogens, such as TGF-β, arrest cell growth by inhibiting cdk activity (9, 53); oncogenic mutations, such as those down-regulating $p16^{Ink4a}$ expression (54, 55) and those up-regulating cyclin D expression (56), render cdks constitutively active. Thus, it is of critical importance to measure cdk activity under various conditions. Cyclin A, E/cdk2 activity is measured using histone H1 as a substrate (Protocol 5), and cyclin D/cdk4,6 activity is measured using pRb as a substrate (Protocol 6).

## Protocol 5

## Histone H1 kinase assays

### Reagents

- NP40-RIPA (see Protocol 4)
- 10 × histone H1 buffer: 200 mM Tris–HCl pH 7.4, 75 mM $MgCl_2$, 10 mM DTT (added just before use)
- 100 mM ATP-lithium salt pH 7.0 (BMB Cat. No. 1140965); stock solution
- 4 mg/ml histone H1 in $H_2O$ (BMB Cat. No. 223549); store at −20 °C

### Method

1. Add 10 μg of antibody against cyclin A, cyclin E, or cdk2 to 200–1000 μg of cellular extract ($5 \times 10^6$ to $5 \times 10^7$ cells) or to a predetermined amount of cyclin A or E/cdk2 baculolysates in 100 μl of NP40-RIPA. Mix and incubate on ice for 1 h.
2. Add 100 μl protein A–Sepharose (or protein G if appropriate for the antibody) slurry equilibrated four times in NP40-RIPA. Rotate at 4 °C for 45 min.
3. Microcentrifuge at 4 °C for 3 sec. Aspirate the supernatant and wash the beads with 1 ml of NP40-RIPA, making sure the beads are completely resuspended. Microcentrifuge again at 4 °C for 3 sec, and wash once more.
4. Wash four times with 900 μl of 1 × H1 kinase buffer. Before aspirating the supernatant of the final wash prepare the reaction mix.
5. For each reaction, mix:
   - 5 ml of 10 × H1 kinase buffer
   - 5 μl of 0.3 mM ATP-lithium salt pH 7.0
   - 0.5 μl of 4 mg/ml histone H1
   - 1 μl of [$\gamma^{32}$P]ATP (3000 Ci/mmol)
   - 38.5 μl $dH_2O$
6. Aspirate the supernatant of the final wash completely with a gel loading tip. Resuspend the pellet in 45 μl of ice-cold reaction mix.
7. Incubate at 37 °C for 30 min. Flick the tubes at 7, 15, and 22 min into the incubation to keep the protein A–Sepharose in suspension.
8. Microcentrifuge at 4 °C for 30 sec.

**Protocol 5** continued

9 Add 15 μl of 4 × SDS–PAGE sample buffer, boil for 5 min, microcentrifuge, and load 20 μl of the supernatant onto a 12% SDS–PAGE gel. Run the gel for 5 min after the dye front runs off. Histone H1 is approx. 34 kDa and runs as a doublet.

10 Stain the gel with Coomassie for 20 min and destain overnight. Dry and expose to X-ray film at −80 °C with an intensifying screen for the required length of time, and/or expose the gel to a phosphorimager screen if the signal is to be quantitated.

# Protocol 6

## Rb kinase assays

### Reagents

- Tween-LB (see Protocol 4)
- 5 × Rb kinase buffer: 250 mM Hepes–KOH pH 7.5, 50 mM $MgCl_2$, 5 mM DTT, 12.5 mM EGTA, 50 mM β-glycerophosphate
- 100 mM ATP-lithium salt pH 7.0 (see Protocol 5)
- GST-Rb or Rb peptide (Santa Cruz, Cat. No. 4112)

### Method

1 Add 10 μg of antibody against cyclin D, cdk4, or cdk6 to 200–1000 μg cellular extract ($5 \times 10^6$ to $5 \times 10^7$ cells) or to a predetermined amount of cyclin D/cdk4 or cdk6 baculolysates in 100 μl of Tween-LB. Mix and incubate on ice for 1 h.

2 Add 100 μl protein A–Sepharose (or protein G if appropriate for the antibody) slurry equilibrated four times in Tween-LB. Rotate at 4 °C for 45 min.

3 Microcentrifuge at 4 °C for 3 sec. Aspirate the supernatant and wash the beads with 1 ml of Tween-LB, making sure the beads are completely resuspended. Microcentrifuge again at 4 °C for 3 sec, and wash three times more.

4 Wash twice with 900 μl of 1 × Rb kinase buffer. Before aspirating the supernatant of the final wash prepare the reaction mix.

5 For each reaction, mix:
   - 6 μl of 5 × kinase buffer
   - 5 μg GST-Rb or 0.5 μl Rb peptide
   - 5 μl of 0.3 mM ATP-lithium salt pH 7.0
   - 1 μl of [$\gamma^{32}$P]ATP
   - $dH_2O$ to 30 μl

6 Aspirate the supernatant of the final wash completely with a gel loading tip. Resuspend the pellet in 30 μl of ice-cold reaction mix.

7 Incubate at 30 °C for 30 min, mixing every 10 min.

8 Microcentrifuge and add 10 μl of 4 × SDS sample buffer. Boil for 5 min, microcentrifuge, and load 20 μl of the supernatant onto a 10% SDS-PAGE gel.

9 Stain the gel with Coomassie for 20 min and destain overnight. Dry and expose to X-ray film at −80 °C with an intensifying screen for the required length of time, and/or expose the gel to a phosphorimager screen if the signal is to be quantitated.

The protocols outlined in this section have been used primarily to study events occurring during G1 phase, or regulating the transitions from G0 to G1 or from G1 to S. Studies of G2/M phase events, such as the mitotic spindle assembly checkpoint and degradation of the APC, have been more successfully carried out using *Xenopus* extracts and oocytes. Cell-free extracts of unfertilized *Xenopus* eggs can replicate chromosomal DNA *in vitro*. When DNA is introduced into these extracts, it is first assembled into chromatin, and then into structures resembling normal interphase nuclei, before it is replicated. This system is therefore also particularly well suited for studies of nuclear formation and its impact on cell cycle control. Instructions for generating and manipulating *Xenopus* extracts and oocytes can be found in ref. 57.

# 5 Conclusions

Using the protocols given in this chapter it is possible to elucidate a novel cell cycle pathway or mode of regulation. Once a cell cycle arrest has been confirmed by PI staining and measurement of [$^3$H]Tdr incorporation, extracts can be made and the amount of cyclin/cdk activity within them can be measured. If necessary, these extracts can be mixed together or complemented with purified recombinant cyclins, cdks, or inhibitors or substrates of these kinases. This paradigm was utilized successfully to identify p27$^{Kip1}$ (9, 10) and other aspects of regulation of the cell cycle (58). Although anything that regulates the cell cycle must ultimately regulate the activity of the cdks, there are many ways for signals to travel from outside of the cell to these nuclear kinases; thus, there may be many modes of regulation left to be discovered applying the tools outlined here.

# Acknowledgements

The authors thank all of the past and present members of the Koff laboratory for sharing their protocols and insights. Work in the investigators' laboratory is supported by funds from the NIH (GM52597), the Memorial Sloan-Kettering Cancer Center Core Grant (CA08748), developmental funding through the SPORE program of NCI (CA68425), and CapCURE. A. K. is a recipient of a Pew Scholarship in Biomedical Science, a Hirschl Scholarship, and is the incumbent of the Frederick R. Adler Chair for Junior Faculty.

# References

1. Sherr, C. J. (1993). *Cell*, **73**, 1059.
2. Lundberg, A. S. and Weinberg, R. A. (1998). *Mol. Cell. Biol.*, **18**, 753.
3. Grana, X., Garriga, J., and Mayol, X. (1998). *Oncogene*, **17**, 3365.
4. Zhang, H. S., Gavin, M., Dahiya, A., Postigo, A. A., Ma, D., Luo, R. X., *et al.* (2000). *Cell*, **101** (1), 79.
5. Park, M. S. and Koff, A. (1998). In *Current protocols in cell biology* (ed. J. S. Bonifacino, M. Dasso, J. B. Harford, J. Lippincott-Schwartz, and K. M. Yamada), p. 8.1.1. John Wiley and Sons, NY.

6. Peterson, B. O., Lukas, J., Sørensen, C. S., Bartek, J., and Helin, K. (1999). *EMBO J.*, **18** (2), 396.
7. Zhao, J., Dynlacht, B., Imai, T., Hori, T., and Harlow, E. (1998). *Genes Dev.*, **12**, 456.
8. Sheaff, R. J., Groudine, M., Gordon, M., Roberts, J. M., and Clurman, B. E. (1997). *Genes Dev.*, **11**, 1464.
9. Koff, A., Ohtsuki, M., Polyak, K., Roberts, J. M., and Massague, J. (1993). *Science*, **260**, 536.
10. Polyak, K., Kato, J. Y., Solomon, M. J., Sherr, C. J., Massague, J., Roberts, J. M., *et al.* (1994). *Genes Dev.*, **8**, 9.
11. Becton Dickinson. (1994). *Becton Dickinson source book*, Section 1.11. Becton Dickinson Immunocytometry Systems, San Jose, CA.
12. Deng, C., Zhang, P., Harper, J. W., Elledge, S. J., and Leder, P. (1995). *Cell*, **82**, 675.
13. Brugarolas, J., Chandrasekaran, C., Gordon, J. I., Beach, D., Jacks, T., and Hannon, G. J. (1995). *Nature*, **377**, 552.
14. (1986). *Cytometry*, **6** (6) – entire issue.
15. Schimenti, K. and Jacobberger, J. W. (1992). *Cytometry*, **13**, 48.
16. Zhang, D. and Jacobberger, J. W. (1998). *Cell Prolif.*, **29**, 289.
17. Sramkowski, R. M., Wormsley, S. W., Bolton, W. E., Crumpler, D. C., and Jacobberger, J. W. (1999). *Cytometry*, **35**, 274.
18. Gray, J. W., Dolbeare, F., Pallavicini, M. G., and Vanderlaan, M. (1987). In *Techniques in cell cycle analysis* (ed. J. W. Gray and Z. Darzynkiewicz), p. 93. Humana Press, Clifton, NJ.
19. Resnitzky, D. and Reed, S. I. (1995). *Mol. Cell. Biol.*, **15** (7), 3463.
20. Geng, Y., Whoriskey, W., Park, M. Y., Bronson, R. T., Medema, R. H., Li, T., *et al.* (1999). *Cell*, **97**, 767.
21. Clarke, A. R., Maandag, E. R., van Roon, M., van der Lugt, N. M., van der Valk, M., Hooper, M. L., *et al.* (1992). *Nature*, **359**, 328.
22. Jacks, T., Fazeli, A., Schmitt, E. M., Bronson, R. T., Goodell, M. A., and Weinberg, R. A. (1992). *Nature*, **359**, 295.
23. Lee, E. Y., Chang, C. Y., Hu, N., Wang, Y. C., Lai, C. C., Herrup, K., *et al.* (1992). *Nature*, **359**, 288.
24. Harrison, D. J., Hooper, M. L., Armstrong, J. F., and Clarke, A. R. (1995). *Oncogene*, **10**, 1615.
25. Hu, N., Gutsmann, A., Herbert, D. C., Bradley, A., Lee, W. H., and Lee, E. Y. (1994). *Oncogene*, **9**, 1021.
26. Cobrinik, D., Lee, M. H., Hannon, G., Mulligan, G., Bronson, R. T., Dyson, N., *et al.* (1996). *Genes Dev.*, **10**, 1633.
27. LeCouter, J. E., Kablar, B., Whyte, P. F. M., Ying, C., and Rudnicki, M. A. (1998). *Development*, **125**, 4669.
28. LeCouter, J. E., Kablar, B., Hardy, W. R., Ying, C., Megeney, L. A., May, L. L., *et al.* (1998). *Mol. Cell. Biol.*, **18** (12), 7455.
29. Lee, M. H., Williams, B. O., Mulligan, G., Mukai, S., Bronson, R. T., Dyson, N., *et al.* (1996). *Genes Dev.*, **10**, 1621.
30. Fero, M. L., Rivkin, M., Tasch, M., Porter, P., Carow, C. E., Firpo, E., *et al.* (1996). *Cell*, **85**, 733.
31. Kiyokawa, H., Kineman, R. D., Manova-Todorova, K. O., Soares, V. C., Hoffman, E. S., Ono, M., *et al.* (1996). *Cell*, **85**, 721.
32. Nakayama, K., Ishida, N., Shirane, M., Inomata, A., Inoue, T., Shishido, N., *et al.* (1996). *Cell*, **85**, 707.
33. Yan, Y., Frisen, J., Lee, M. H., Massague, J., and Barbacid, M. (1997). *Genes Dev.*, **11**, 973.
34. Zhang, P., Liegeois, N. J., Wong, C., Finegold, M., Hou, H., Thompson, J. C., *et al.* (1997). *Nature*, **387**, 151.

35. Zhang, P., Wong, C., Liu, D., Finegold, M., Harper, J. W., and Elledge, S. J. (1999). *Genes Dev.*, **13**, 213.
36. Zhang, P., Wong, C., DePinho, R. A., Harper, J. W., and Elledge, S. J. (1998). *Genes Dev.*, **12**, 3162.
37. Serrano, M., Lee, H., Chin, L., Cordon-Cardo, C., Beach, D., and DePinho, R. A. (1996). *Cell*, **85**, 27.
38. Roussel, M. F. (1999). *Oncogene*, **18**, 5311.
39. Franklin, D. S., Godfrey, V. L., Lee, H., Kovalev, G. I., Schoonhoven, R., Chen-Kiang, S., *et al.* (1998). *Genes Dev.*, **12**, 2899.
40. Zindy, F., van Deursen, J., Grosveld, G., Sherr, C. J., and Roussel, M. F. (2000). *Mol. Cell. Biol.*, **20**, 372.
41. Latres, E., Malumbres, M., Sotillo, R., Martín, J., Ortega, S., Martín-Caballero, J., *et al.* (2000). *EMBO J.*, **19** (13), 3496.
42. Zindy, F., Cunningham, J. J., Sherr, C. J., Jogal, S., Smeyne, R. J., and Roussel, M. F. (1999). *Proc. Natl. Acad. Sci. USA*, **96**, 13462.
43. Park, M. S., Rosai, J., Nguyen, H. T., Capodieci, P., CordÓn-Cardo, C., and Koff, A. (1999). *Proc. Natl. Acad. Sci. USA*, **96**, 6382.
44. Coats, S., Whyte, P., Fero, M. L., Lacy, S., Chung, G., Randel, E., *et al.* (1999). *Curr. Biol.*, **9**, 163.
45. Sicinski, P., Donaher, J. L., Parker, S. B., Li, T., Fazeli, A., Gardner, H., *et al.* (1995). *Cell*, **82**, 621.
46. Sicinski, P., Donaher, J. L., Geng, Y., Parker, S. B., Gardner, H., Park, M. Y., *et al.* (1996). *Nature*, **384**, 470.
47. Rane, S. G., Dubus, P., Mettus, R. V., Galbreath, E. J., Boden, G., Reddy, E. P., *et al.* (1999). *Nature Genet.*, **22**, 44.
48. Tsutsui, T., Hesabi, B., Moons, D. S., Pandolfi, P. P., Hansel, K. S., Koff, A., *et al.* (1999). *Mol. Cell. Biol.*, **19**, 7011.
49. LaBaer, J., Garrett, M. D., Stevenson, L. F., Slingerland, J. M., Sandhu, C., Chou, H. S., *et al.* (1997). *Genes Dev.*, **11**, 847.
50. Cheng, M., Olivier, P., Diehl, J. A., Fero, M., Roussel, M. F., Roberts, J. M., *et al.* (1999). *EMBO J.*, **18**, 1571.
51. Desai, D., Gu, Y., and Morgan, D. O. (1992). *Mol. Biol. Cell*, **3**, 571.
52. Kato, J. Y., Matsuoka, M., Strom, D. K., and Sherr, C. J. (1994). *Mol. Cell. Biol.*, **14** (4), 2713.
53. Hannon, G. J. and Beach, D. (1994). *Nature*, **371**, 257.
54. Kamb, A., Shattuck-Eidens, D., Eeles, R., Liu, Q., Gruis, N. A., Ding, W., *et al.* (1994). *Nature Genet.*, **8**, 23.
55. Ruas, M. and Peters, G. (1998). *Biochim. Biophys. Acta*, **1378**, F115.
56. Stteg, P. S. and Zhou, Q. (1998). *Breast Cancer Res. Treat.*, **52** (1–3), 17.
57. Dunphy, W. G. (ed.) (1997). *Methods in enzymology*, Vol. 283, p. 535. Academic Press, NY.
58. Gitig, D. M. and Koff, A. (2000). In *Transforming growth factor-beta protocols* (ed. P. H. Howe), p. 109. Humana Press Inc., Totowa, NJ.
59. Vidal, A. and Koff, A. (2000). *Gene*, **247**, 1.

# Chapter 4
# Cellular bioenergetics

David G. Nicholls
Buck Institute for Age Research, Novato, California, USA.

Manus W. Ward
Mitokor, San Diego, California, USA.

## 1 Introduction

Mitochondria perform multiple functions within cells. In addition to their primary role in the generation of ATP for cytoplasmic and plasma membrane energy-requiring processes, there is an increasing realization that mitochondria can play a central role in the regulation of cytoplasmic calcium concentrations, while the mitochondrial respiratory chain is the major source of potentially toxic reactive oxygen species (1). The chemiosmotic mitochondrial proton circuit (Figure 1) which links electron transport through the respiratory chain to the ATP synthase can be considered in many ways analogous to an equivalent electrical circuit, with potential, flux, and resistance elements obeying Ohm's law. The potential term is the protonmotive force, $\Delta p$, amounting to some 200 mV and comprised of membrane potential, $\Delta\psi_m$, and $\Delta pH$ components according to the equation:

$$\Delta p = \Delta\psi_m - 60\Delta pH$$

While $\Delta p$ drives ATP synthesis, it is $\Delta\psi_m$ that is responsible for $Ca^{2+}$ accumulation, while the magnitude of the membrane potential also largely controls the rate at which mitochondria generate reactive oxygen species (2). In mitochondria possessing spare respiratory capacity, respiration is controlled by proton re-entry into the matrix via the ATP synthase plus endogenous proton leak pathways. This is signalled to the respiratory chain as a slight drop in $\Delta p$. Techniques have evolved over many years that have allowed the key bioenergetic parameters to be quantified for isolated mitochondria and in this chapter we shall discuss the extension of these techniques to the investigation of mitochondria *in situ* within cultured cells.

## 2 Mitochondrial membrane potential

The membrane potential and pH components of the protonmotive force have been determined separately, from the equilibrium distribution of lipophilic,

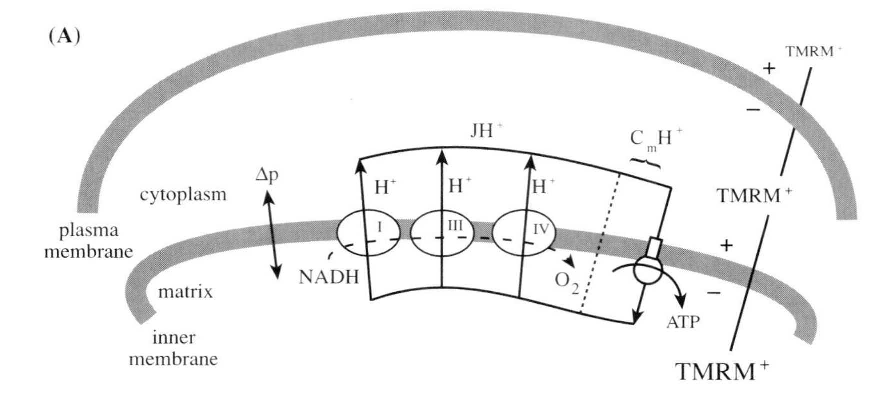


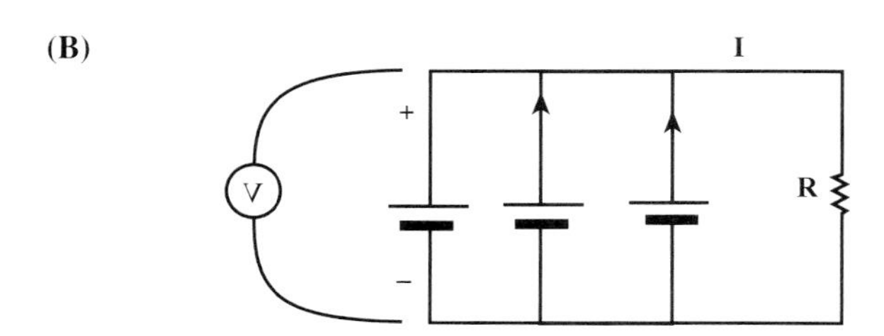


**Figure 1** The mitochondrial proton circuit. (A) Complexes I, III, and IV function as proton pumps, in parallel with respect to the proton circuit and in series with respect to electron transport. The proton circuit is completed by proton re-entry via the ATP synthase or via an endogenous proton leak (dotted line). A lipophilic cation such as $TMRM^+$ is accumulated to a Nernst equilibrium across both the plasma and inner mitochondrial membranes. The proton circuit is analogous to an equivalent electrical circuit, (B), with the protonmotive force, Δp equivalent to the voltage term, V; the proton current $JH^+$ equivalent to the electrical current, I, and the two related to the effective proton conductance, $C_mH^+$, or resistance, R, by Ohm's law.

membrane permeant cations and weak acids or bases respectively. In the presence of excess phosphate, the ΔpH component of Δp is small, and changes in membrane potential can generally be assumed to reflect parallel changes in the total protonmotive force. For that reason, the vast majority of studies with isolated mitochondria focus purely on the membrane potential.

Lipophilic cations such as tetraphenylphosphonium, $TPP^+$, distribute across the inner membrane to a Nernst equilibrium, after correcting for binding, and are thus accumulated within the mitochondrial matrix by ten-fold for every 60 millivolts of membrane potential. The accumulation can be quantified either by utilizing radiolabelled cation and rapidly separating mitochondria from medium (3, 4), or by taking advantage of a $TPP^+$ electrode to measure the residual cation concentration in the external medium (5). This latter has the advantage that changes in $\Delta\psi_m$ can be monitored continuously.

More qualitative monitoring of mitochondrial membrane potential can be achieved by the use of fluorescent membrane permeant cations, used under conditions when the high accumulation of the probe within the mitochondrial matrix leads to stacking and consequent fluorescent quenching. Under these conditions, the total fluorescence of a stirred suspension of mitochondria within the cuvette will decrease as the organelles accumulate probe. This technique is, however, largely empirical, and care must be taken to ensure that the high accumulation of probe within the mitochondrial matrix does not affect mitochondrial function.

The extension of these techniques to monitor the membrane potential of mitochondria *in situ* within functional cells is fraught with additional complications, since the mitochondria are not directly accessible, residing within a cytoplasm which is in turn bounded by a plasma membrane. The equilibrium accumulation of the cation within the mitochondrial matrix will thus be dependent not only on $\Delta\psi_m$ but also on the plasma membrane potential. Estimations of *in situ* mitochondrial membrane potential using phosphonium cations have to take into account the plasma membrane potential, the relative volume of mitochondrial matrix and cytoplasm, and the extent of binding in the two compartments. When this is done, the best estimates for the *in situ* mitochondrial membrane potential under respiring conditions are close to 150 millivolts (3, 4).

In this chapter we focus on the use of fluorescent probes to monitor *in situ* mitochondrial function. Two levels of resolution can be achieved. Conventional fluorescence microscopy accompanied by digital imaging cannot readily resolve single mitochondria within the cell. Instead, the total cellular fluorescence has to be processed to extract information on the mitochondrial potential. Two distinctive methods are required to monitor, on the one hand, relatively rapid changes in $\Delta\psi_m$ during the period of observation, and, on the other, to detect changes in the distribution of potentials in a cell population that have already occurred as result of experimental manipulations prior to the start of the period of observation.

## 2.1 Monitoring dynamic changes in mitochondrial membrane potential with $TMRM^+$ or rhodamine 123

Tetramethylrhodamine methyl ester ($TMRM^+$) and the closely related ethyl ester, $TMRE^+$, are highly permeant potentiometric probes (6) which may be used in either matrix-quenching or non-quenching modes, and are thus suitable for monitoring rapid dynamic changes in potential at single cell resolution under conditions of matrix quench. At low, non-quenching concentrations the probe can monitor equilibrium mitochondrial potential at single cell resolution or even single mitochondrial potentials in the confocal microscope.

In order to monitor dynamic changes in mitochondrial membrane potential, the probe loading has to be sufficient to exceed the threshold for quenching within the matrix. In practice, for a cultured neuron, this means equilibrating the cells with at least 50 nM $TMRM^+$. As with all probes, the concentration employed should be the minimum which satisfies this criterion. $TMRM^+$ is rather less toxic to mitochondrial function than most other cationic probes (7), but care has to be taken to ensure that the probe is not modifying mitochondrial function. Some cationic cyanine probes are so toxic that they must be used at <1 nM (8). Methods for monitoring changes in mitochondrial membrane potential with $TMRM^+$ or rhodamine 123 are described in Protocol 1.

Because $TMRM^+$ is so permeant it will re-equilibrate across the plasma membrane within the time-course of most experiments, particularly if the cells examined are small or have neurite extensions. It is important to recognize when a change in whole cell fluorescence is a consequence of such distribution, rather than a primary change in $\Delta\psi_m$. Techniques for analysing whole cell fluorescence changes are described elsewhere (9, 10).

Rhodamine 123 is closely related to $TMRM^+$, but is less hydrophobic and consequently permeates some 20-fold more slowly across membranes (11). Rhodamine 123 is loaded by exposure to high concentrations of probe for brief periods insufficient for equilibration. It is then usually washed away and experiments performed in the absence of external probe. The relatively low permeability of rhodamine 123 allows plasma membrane redistribution to be ignored in most short-term experiments (Figure 2).

## Protocol 1

## Monitoring changes in mitochondrial membrane potential with $TMRM^+$ or rhodamine 123

### Equipment and reagents

- Digital imaging facility with inverted microscope, ×40 oil immersion objective
- Single excitation, either 485 nm with emission >510 nm ($TMRM^+$ or rhodamine 123) *or* excitation 535 nm with emission >550 nm ($TMRM^+$)
- $TMRM^+$ stock: 10 μM in DMSO
- Incubation medium: 120 mM NaCl, 3.1 mM KCl, 0.4 mM $KH_2PO_4$, 5 mM $NaHCO_3$, 1.2 mM $Na_2SO_4$, 1.2 mM $MgCl_2$, 1.3 mM $CaCl_2$, 15 mM glucose, and 20 mM TES pH adjusted to 7.4 at 37 °C with NaOH
- Rhodamine 123 stock: 260 μM (100 μg/ml) in water

### A. Method for $TMRM^+$

1. Add 1.5 ml of incubation medium to a culture dish. To this, add 15 μl $TMRM^+$ stock (100 nM final concentration).
2. Place a coverslip of cells into the culture dish and incubate in the dark at 37 °C for 30 min.
3. After the 30 min incubation wash the cells with fresh incubation medium.
4. Incubate the cells in the imager in medium containing 100 nM $TMRM^+$.
5. Monitor at 535 nm excitation and >550 nm emission for maximum sensitivity, or 485 nm excitation, >510 nm emission to reduce photoinduced damage at the concentrations of $TMRM^+$ required to exceed the matrix quench threshold.

**Protocol 1** continued

**B. Method for rhodamine 123**

1. Add 1.5 ml of incubation medium to a culture dish. To this, add 15 μl rhodamine 123 stock (final conc. 2.6 μM).
2. Place a coverslip of cells into the culture dish and incubate in the dark at room temperature for 15 min.
3. After the 30 min incubation wash the cells with fresh incubation medium.
4. Incubate the cells in the imager in medium in the absence of external rhodamine 123.
5. Monitor at 485 nm excitation, >510 nm emission.

## 2.2 Interpretation of dynamic whole cell fluorescence traces obtained with $TMRM^+$ or rhodamine 123

During maintained matrix quenching, the fluorescence signal from the *in situ* mitochondria is essentially potential independent until $\Delta\psi_m$ drops sufficiently to lower the matrix concentration below the quench threshold. How then can this condition be used to monitor dynamic changes in $\Delta\psi_m$? The answer is that the changing whole cell fluorescence signal originates from the cytoplasm, where mitochondrial depolarization results in a temporary excess of highly fluorescent probe released from the quenched environment of the matrix. Even if the plasma membrane potential does not change, this excess probe will subsequently re-equilibrate to restore the Nernst equilibrium across the plasma membrane. It follows that an increased whole cell signal due to dequenching of probe depends on the more rapid equilibration across the small, deeply invaginated inner mitochondrial membrane, compared with that across the large, roughly spherical plasma membrane.

Figure 2 shows a gallery of responses simulated with a simple programme which takes account of matrix quenching and the more rapid approach to Nernst equilibrium across the mitochondrial rather than plasma membranes (10). These simulated responses correspond closely to those obtained experimentally with single cell imaging of small neurons.

## 2.3 Equilibrium monitoring of mitochondrial membrane potential

The dynamic response to changes in $\Delta\psi_m$ described above will fail to detect either very slow changes, or pre-existing differences in potential which might have occurred before the start of the experiment, for example, during prolonged apoptosis experiments or for analysis by fluorescence-activated cell sorting (FACS). Under these conditions, it is essential that the loading concentration of probe is below that required for matrix quenching, such that the equilibrium single cell fluorescence corresponds to the total accumulation of probe within the cell (Figure 2C). It is important to appreciate that, under these conditions, the signal is roughly proportional to the sum of the plasma and mitochondrial membrane potentials (1) and that, if the quench threshold is exceeded, the signal remains sensitive to plasma membrane potential but loses $\Delta\psi_m$ sensitivity

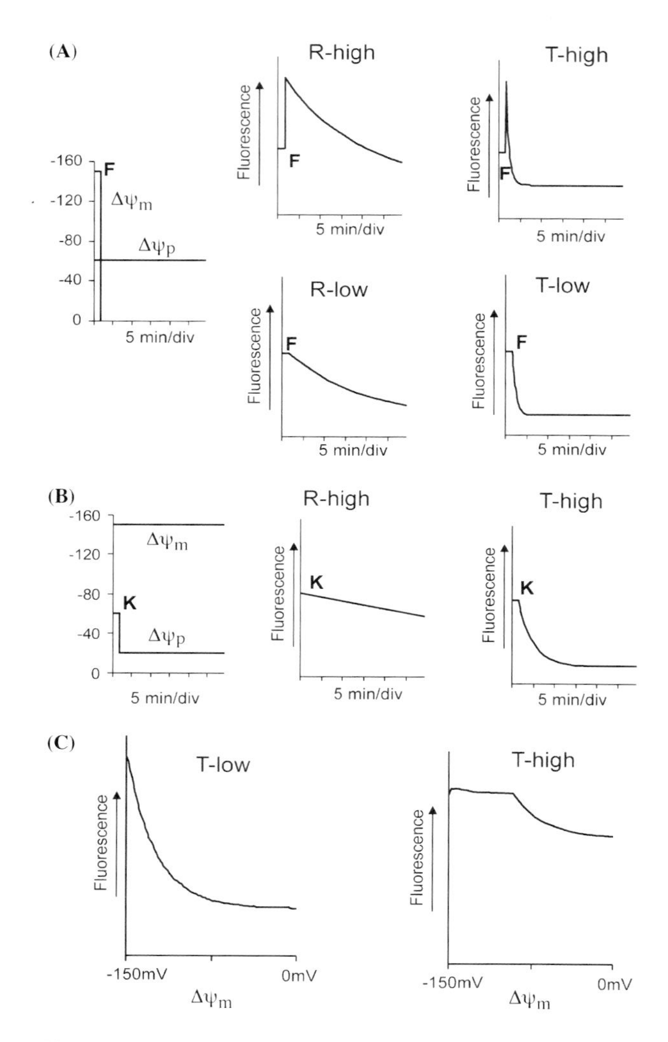


**Figure 2** (A, B) Simulated *dynamic* single cell fluorescence traces for the cell body of a small neuron equilibrated with either high (sufficient for matrix quenching) or low (sub-quenching) concentrations of a fast responding cationic fluorescent probe such as $TMRM^+$ (T), or a slow responding probe such as rhodamine 123 (R). Responses are simulated to (A), a sudden collapse of $\Delta\psi_m$, for example by addition of a protonophore such as FCCP (F) and (B), plasma membrane depolarization, for example by addition of high KCl (K). Note that the protonophore-induced increase in fluorescence is dependent upon initial probe quenching in the matrix. (C) Simulated *equilibrium* single cell fluorescence of a cell equilibrated with a low (just sub-quenching) or high (ten times higher) concentration of a cationic probe as a function of the value of $\Delta\psi_m$ prior to probe addition. Note that sensitivity to $\Delta\psi_m$ is lost above the quench limit.

(8). Failure to appreciate this has led to several erroneous conclusions in apoptotic studies with lymphocytes.

## 2.4 Cytoplasmic free $Ca^{2+}$ concentration with fura-2

Since this is such a standard technique, the conditions used in our laboratory are described briefly in Protocol 2, as an introduction to the simultaneous determination of cytoplasmic $Ca^{2+}$ and $\Delta\psi_m$. One problem with the membrane permeant acetoxymethyl ester form of fura-2 and related probes is its limited solubility in the medium. Rather than use a detergent such as pluronic acid, we use a low concentration of albumin to hold the ester in solution during the loading. Identical experimental conditions are used for the low affinity fura-2FF.

## Protocol 2

### Monitoring cytoplasmic free $Ca^{2+}$ concentrations in cultured neurons with fura-2

#### Equipment and reagents

- Digital imaging facility with inverted microscope, ×40 oil immersion objective
- Dual excitation 340/380 nm, emission >510 nm
- Incubation medium (see Protocol 1)
- Fura-2AM stock: 1 mM in DMSO
- Bovine serum albumin stock: 10 mg/ml in water
- Ionomycin stock: 1 mM in ethanol
- 0.5 M Na-EGTA pH 7.4

#### Method

1. Add 1.5 ml of incubation medium to a culture dish at 37 °C. To this, add 5 μl BSA stock and 5 μl fura-2AM stock. Incubate the dish in the dark at 37 °C for 30 min.
2. Wash the cells with fresh incubation medium.
3. Image with alternate 340/380 nm excitation, emission >505 nm, focusing on neurites or somata as appropriate.
4. At the end of each experiment, calibrate by the sequential addition of 10 μM ionomycin and 10 mM Na-EGTA.

## 2.5 Simultaneous monitoring with $TMRM^+$ or rhodamine 123 of dynamic changes in $\Delta\psi_m$ and cytoplasmic free calcium concentration

In cell death studies it is frequently useful to correlate changes in mitochondrial membrane potential and changes in cytoplasmic calcium homeostasis. As described in Protocol 3, the cells, initially loaded with fura-2AM together with either $TMRM^+$ or rhodamine 123 are washed to remove excess calcium indicator and then monitored in the presence ($TMRM^+$) or absence (rhodamine 123) of external potentiometric probe. Two filter combinations may be used to allow simultaneous monitoring of fura-2 and potential probe fluorescence, excitation at 340 nm plus 380 nm for fura-2, accompanied by 535 nm excitation at the $TMRM^+$ fluorescence peak, allows optimal detection efficiency. However, at the concentrations of probe which must be employed to exceed the matrix quenching threshold, there is a risk of photoinduced damage. Additionally, a special dichroic is required to allow detection of the fura-2 emission without interference from the $TMRM^+$ excitation. An alternative strategy is to excite the TMRM at about 490 nm, where absorbance is 10% of maximum, utilizing a simple dichroic transmitting above 510 nm. This combination is also suitable for fura-2/rhodamine 123.

## Protocol 3

## Measuring $Ca^{2+}$ fluxes and mitochondrial membrane potential simultaneously in cerebellar granule neurons with fura-2AM and $TMRM^+$ or rhodamine 123

### Equipment and reagents

- Digital imaging facility with inverted microscope, ×40 oil immersion objective
- Incubation medium and stock dyes (see Protocols 1 and 2)
- Triple excitation, either 340/380/485 nm with emission >510 nm ($TMRM^+$ or rhodamine 123); *or* excitation 340/380/548 nm with Chroma technology 'fura-2/rhodamine' emission set C – 00908 ($TMRM^+$)

### A. Method for $TMRM^+$

1. Add 1.5 ml of incubation medium to a culture dish. To this, add 5 μl BSA stock and 5 μl fura-2AM stock (final concentration 3 μM) and 15 μl $TMRM^+$ stock (final concentration 100 nM).
2. Place a coverslip of cells into the culture dish and incubate in the dark at 37 °C for 30 min.
3. Wash the cells with fresh incubation medium.
4. Incubate the cells in the imager in medium containing 100 nM $TMRM^+$.

### B. Method for rhodamine 123

1. Add 1.5 ml of incubation medium to a culture dish. To this, add 5 μl BSA stock and 5 μl fura-2AM stock (final concentration 3 μM) and 15 μl rhodamine 123 stock (final concentration 2.6 μM).
2. Place a coverslip of cells into the culture dish and incubate in the dark at 22 °C for 15 min.
3. Wash the cells with fresh incubation medium.
4. Incubate the cells in the imager in medium in the absence of rhodamine 123.

## 3 Monitoring mitochondrial superoxide generation

Under conditions of high mitochondrial membrane potential and/or matrix calcium loading, the mitochondrial respiratory chain generates increased amounts of potentially toxic superoxide free radicals (2). Superoxide anions oxidize hydroethidine to the fluorescent ethidium, and the rate of increase in whole cell fluorescence can be used as a monitor of the concentrations of superoxide existing within the mitochondrial matrix. Hydroethidine is a cell permeant, blue fluorescent dye until oxidized to ethidium, which has an excitation maximum 495 nm and emission maximum 600 nm. Hydroethidine oxidation is relatively specific for superoxide (12). In contrast to dichlorofluorescein (see below), ethidium fluorescence is relatively photostable under normal illumination conditions and auto-oxidation is negligible for at least 30 min. However, slow auto-oxidation of hydroethidine to ethidium occurs in concentrated stocks stored in either methanol or DMSO under nitrogen at −20 °C.

One complication with this technique, described in Protocol 4, is that the fluorescent product is a membrane permeant cation and can therefore redistribute between matrix and cytoplasm in response to change in $\Delta\psi_m$. Ethidium intercalates into both mitochondrial and nuclear DNA with considerable fluorescent enhancement, and it is essential to use very low concentrations of hydroethidine so that generated ethidium remains largely bound to mitochondrial DNA, rather than remaining free so that a subsequent mitochondrial depolarization would lead to its efflux, binding to nuclear DNA, and consequent fluorescent enhancement. Failure to take this precaution has led to suggestions that mitochondrial depolarization enhances the production of superoxide within cells, contradicting results obtained with isolated mitochondria.

### Protocol 4

### Superoxide detection with hydroethidine

#### Equipment and reagents

- Digital imaging facility with inverted microscope, ×40 oil immersion objective
- Incubation medium (see Protocol 1)
- Single excitation, 485 nm with emission >510 nm

#### Method

1. Prepare a fresh stock of 2 mM hydroethidine in 100% dimethyl sulfoxide each day. Dilute before use to 200 μM in nitrogen-purged water.
2. Add Dowex-50W-X8 cationic resin beads to the tube and vortex remove contaminant ethidium cation. Keep on ice until use.
3. Add 1 μM hydroethidine to cells in imager without pre-incubation. Excite at 385 nm and monitor emission at >520 nm. It is important not to exceed this loading concentration since the whole cell fluorescence of ethidium can then become dependent on mitochondrial membrane potential (13).

Hydroethidine cannot be used in combination with UV-excited probes such as fura-2, since the 340 nm illumination causes a rapid photo-oxidation of the probe (12).

## 4 Monitoring hydrogen peroxide generation

Dichlorofluorescin (DCF-$H_2$) is commonly employed to monitor the production of hydrogen peroxide within cells. A suitable method for this is described in Protocol 5. This probe has the disadvantage that its fluorescence is sensitive to changes in cytoplasmic pH within the range that can occur during physiological manipulations. The esterified form of dichlorofluorescin, dichlorofluorescin diacetate, readily crosses cell membranes and then undergoes deacetylation by intracellular esterases. The resulting non-fluorescent dichlorofluorescin is trapped within the cell and susceptible to reactive oxygen species-mediated oxidation to the fluorescent dichlorofluorescein. It has been reported that DCF-$H_2$ is oxidized most readily by hydrogen peroxide but not by superoxide (12).

### Protocol 5

### Hydrogen peroxide measurement with 2′,7′-dichlorofluorescin (DCF-$H_2$)

#### Equipment and reagents

- Digital imaging facility with inverted microscope, ×40 oil immersion objective
- Single excitation, 485 nm with emission >510 nm
- Incubation medium (see Protocol 1)
- 1 mM DCF-$H_2$ in DMSO (stock solution); make as a concentrated stock each day and store under nitrogen between experiments at −20 °C

#### Method

1. Load coverslip-mounted cells with 10 μM DCF-$H_2$ for 10 min at 37 °C in the dark.
2. Rinse cells once after loading and then mount in imager. Monitor DCF fluorescence in individual cells by excitation at 495 nm and emission >505 nm. Once determined, the optical parameters (exposure etc.) for illumination must remain fixed for the experiment.
3. Calibrate intracellular dichlorofluorescein fluorescence using 0.2–30 mM hydrogen peroxide.

## 5 ATP/ADP ratios in cultured neurons

The ratio of ATP to ADP within the cell population provides a monitor of bioenergetic status that is more useful than that obtained by ATP analysis alone. ATP is determined rapidly in cell extracts by conventional luciferase assay, following which ADP in the extract is converted to ATP by pyruvate kinase in the presence of phospho*enol*pyruvate. A method for determining the ATP/ADP ratios in cultured neurons is described in Protocol 6.

## Protocol 6

## ATP/ADP ratios in populations of cultured neurons

### Equipment and reagents

- LKB Wallac 1250 tube luminometer or equivalent
- Extraction medium: 1 M perchloric acid/50 mM EDTA at 0 °C
- Neutralization medium: 3 M KOH in 1.5 M Tris base
- Chemiluminescence medium: firefly luciferase and D-luciferin (Labtech International, 1:5 final dilution) in 0.1 mM Tris–acetate, 2 mM EDTA pH 7.75

### Method

1. Extract adenine nucleotides from cells by the addition of 100 μl extraction medium at 0 °C.
2. Centrifuge extracts in a microcentrifuge, neutralize with neutralization medium, cool to 0 °C, and then re-centrifuge.
3. Add aliquots of the supernatant immediately to a luminometer assay vial containing 200 μl chemiluminescence medium. The increase in chemiluminescence is proportional to the concentration of ATP in the aliquot.
4. When a stable response is obtained, add pyruvate kinase (2 U per assay) and 0.5 mM phospho*enol*pyruvate (PEP), and determine the further increase in chemiluminescence due to the conversion of ADP to ATP.
5. Add the ATP standard to calibrate the luminescence reaction.

# References


1. Nicholls, D. G. and Budd, S. L. (2000). *Physiol. Rev.*, **80**, 315.
2. Skulachev, V. P. (1996). *Q. Rev. Biophys.*, **29**, 169.
3. Hoek, J. B., Nicholls, D. G., and Williamson, J. R. (1980). *J. Biol. Chem.*, **255**, 1458.
4. Scott, I. D. and Nicholls, D. G. (1980). *Biochem. J.*, **186**, 21.
5. Kamo, N., Muratsugu, M., Hongoh, R., and Kobatake, Y. (1979). *J. Membr. Biol.*, **49**, 105.
6. Ehrenberg, B., Montana, V., Wei, M. D., Wuskell, J. P., and Loew, L. M. (1988). *Biophys. J.*, **53**, 785.
7. Scaduto, R. C. and Grotyohann, L. W. (1999). *Biophys. J.*, **76**, 469.
8. Rottenberg, H. and Wu, S. L. (1998). *Biochim. Biophys. Acta*, **1404**, 393.
9. Nicholls, D. G. and Ward, M. W. (2000). *Trends Neurosci.*, **23**, 166.
10. Ward, M. W., Rego, A. C., Frenguelli, B. G., and Nicholls, D. G. (2000). *J. Neurosci.*, **20**, 7208.
11. Bunting, J. R. (1992). *Biophys. Chem.*, **42**, 163.
12. Bindokas, V. P., Jordan, J., Lee, C. C., and Miller, R. J. (1996). *J. Neurosci.*, **16**, 1324.
13. Budd, S. L., Castilho, R. F., and Nicholls, D. G. (1997). *FEBS Lett.*, **415**, 21.

# Chapter 5

# Targeting of nuclear-encoded proteins into and across the thylakoid membrane: isolation and analysis of intact chloroplasts and thylakoids from plants

Colin Robinson
Department of Biological Sciences, University of Warwick, Coventry CV4 7AL, UK.

Alexandra Mant
Plant Biochemistry Laboratory, Department of Plant Biology, The Royal Veterinary and Agricultural University, 1871 Frederiksberg C, Copenhagen, Denmark.

## 1 Introduction

The biogenesis of the chloroplast is a complex process, involving a great deal of protein traffic. The majority of chloroplast proteins (around 80%) are imported from the cytosol, because they are encoded by nuclear genes (reviewed in refs 1 and 2). These proteins must be both specifically targeted into the chloroplast and accurately sorted to their final sites of function within the organelle. With the exception of a subset of outer envelope proteins, which have no cleavable targeting peptides (3), all imported proteins analysed to date are initially synthesized as larger precursors containing amino terminal presequences. The presequences have been shown to contain information which specifies targeting into the chloroplast, and in some cases, localization within the organelle.

This sorting of imported proteins is particularly interesting, due to the structural complexity of the chloroplast, which consists of three membrane compartments (outer and inner envelopes and the thylakoid membrane) and three soluble compartments (the intermembrane space, the stroma, and the thylakoid lumen). Each of these compartments contains cytosolically synthesized proteins, and intensive efforts have been made to understand the mechanisms which ensure correct targeting and intra-organellar routing of imported proteins. These studies have depended totally on the development of efficient *in vitro* assays for protein translocation across the envelope and thylakoid membranes. This article describes in detail basic protocols for the import of *in vitro* synthesized proteins into both intact chloroplasts and isolated thylakoids.

## 1.1 Choice of plants

The correct choice of plant is of great importance—only a few higher plant species yield intact chloroplasts capable of efficient protein import. The majority of species contain phenolic compounds in the vacuole, which become oxidized upon disruption of the tissue and are believed to inactivate isolated organelles. Seedlings from dwarf pea varieties (*Pisum sativum*) most commonly provide the starting material for the isolation of intact chloroplasts. Pea chloroplasts are easily and rapidly isolated, and will happily import precursor proteins from a variety of monocot and dicot species, although the uptake efficiency can vary between proteins. In our laboratory we currently use 'Kelvedon Wonder', which may be purchased from Nickerson-Zwaan, although for many years we used 'Feltham First' which, at the time of writing, is unavailable from UK seed merchants.

Spinach (*Spinacia oleracea*) is another abundant source of suitable chloroplasts. However, it has two disadvantages compared to pea: first it grows more slowly and secondly, the isolated thylakoid membranes do not import/insert proteins so easily.

Recently *Arabidopsis thaliana* chloroplasts have been persuaded to import proteins—descriptions of growth conditions and isolation methods may be found in refs 4 and 5. Overall however, it appears that *A. thaliana* chloroplasts import proteins less efficiently than pea, and we are not yet aware of any reports of protein uptake using isolated thylakoid membranes.

Wheat (*Triticum aestivum*) (6), maize (*Zea mays*) (7), and barley (*Hordeum vulgare*) (8) will also yield import-competent chloroplasts although again, isolation is slightly trickier than for pea.

## 1.2 Growth conditions

Peas may be planted in compost (e.g. Levington's multipurpose compost) or vermiculite. Water particularly well on the first day, to allow the seeds to speed imbibition and germination. We grow the seedlings for seven to ten days at 18–25 °C under a 12 hour photoperiod. The light intensity should be relatively low (40–50 $\mu E\ m^{-2}\ s^{-1}$) to minimize the formation of starch grains which cause chloroplast rupture during the isolation procedure. As an extra precaution against excessive starch deposits, pea leaves should ideally be harvested about an hour after 'sun-up'. Only young leaves should be used, as isolated chloroplasts from mature tissue are no longer import-competent (6). Figure 1 shows pea seedlings at the perfect stage for chloroplast isolation, around two to three days after the emergence of the first leaves. Two trays of seedlings should be more than sufficient for an experiment (expect a total yield of 1–5 mg chlorophyll).

# 2 Isolation of intact chloroplasts

There are several important considerations to bear in mind when isolating chloroplasts. As soon as the tissue has been homogenized, the protocol should

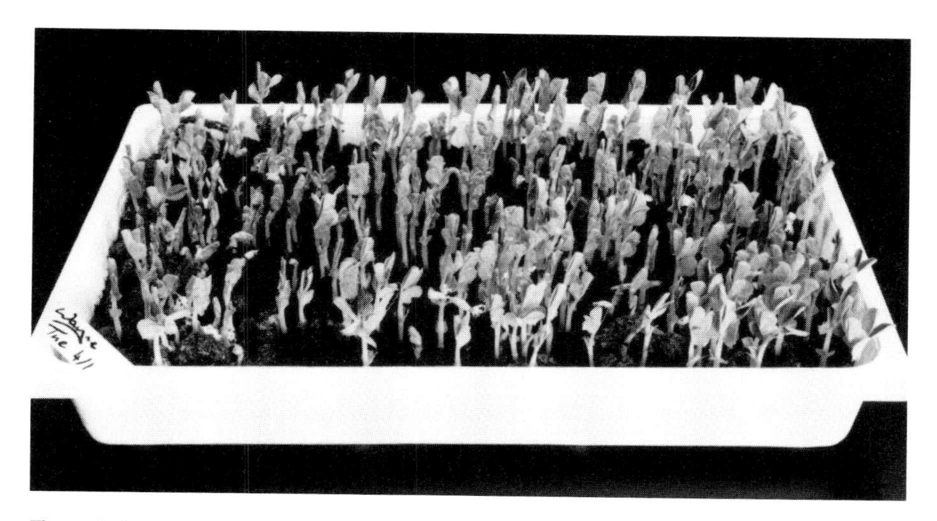

**Figure 1** Growth of pea seedlings for chloroplast isolation. This seed tray contains pea seedlings (var. Kelvedon Wonder) at the best growth stage for isolation of import-competent chloroplasts, which is two to three days after leaf emergence. Under our greenhouse conditions, the seedlings are eight to nine days old. At the bottom right-hand corner of the tray are a few seedlings whose leaves have fully opened—these ones are a little bit over-grown.

be worked through rapidly, as the chloroplasts' ability to import proteins decreases over time. All the materials used should be ice-cold, including of course, the chloroplast suspension. If a refrigerated centrifuge is unavailable, just make sure that the centrifuge buckets themselves have been refrigerated as the centrifugation times involved are not very long. Most importantly, the grinding medium should be placed in a freezer until it forms an icy slurry (shaking the bottle of super-cooled buffer gives a very satisfying instant slush). The heat energy created by the homogenization process is absorbed by the slurry as it melts. Our homogenizer is a Polytron fitted with a PTA 20 TSM aggregate (one of many types), from Kinematica AG. However, other types of blender will serve just as well, as long as the grinding blades are very sharp, frothing is kept to a minimum, and the overall homogenization time is short. Indeed, the most efficient blender ever used by the authors was a domestic food processor customized so that the rotor carried disposable razor blades, which were changed after every chloroplast preparation. If at all possible, reserve a set of centrifuge tubes to be used only for chloroplast isolation, and make sure these are cleaned gently to avoid scratches, and without the use of any sort of detergent, as scratched tubes and detergent residue will both cause lysis of the organelles. Finally, great care must be taken in handling chloroplasts; their large size means that they are prone to lysis from shearing forces, such as those caused by pipetting. When it is necessary to pipette a suspension of chloroplasts, widen the bore of disposable tips by cutting off the ends.

Original protocols for the isolation of pea chloroplasts contained more washing steps, and the buffers were more complex. Over the years however, we have gradually simplified the isolation procedure for **pea** chloroplasts without any loss in the yield or quality of the intact organelles. Protocol 1 is, therefore, a streamlined method which will save time and consumables money.

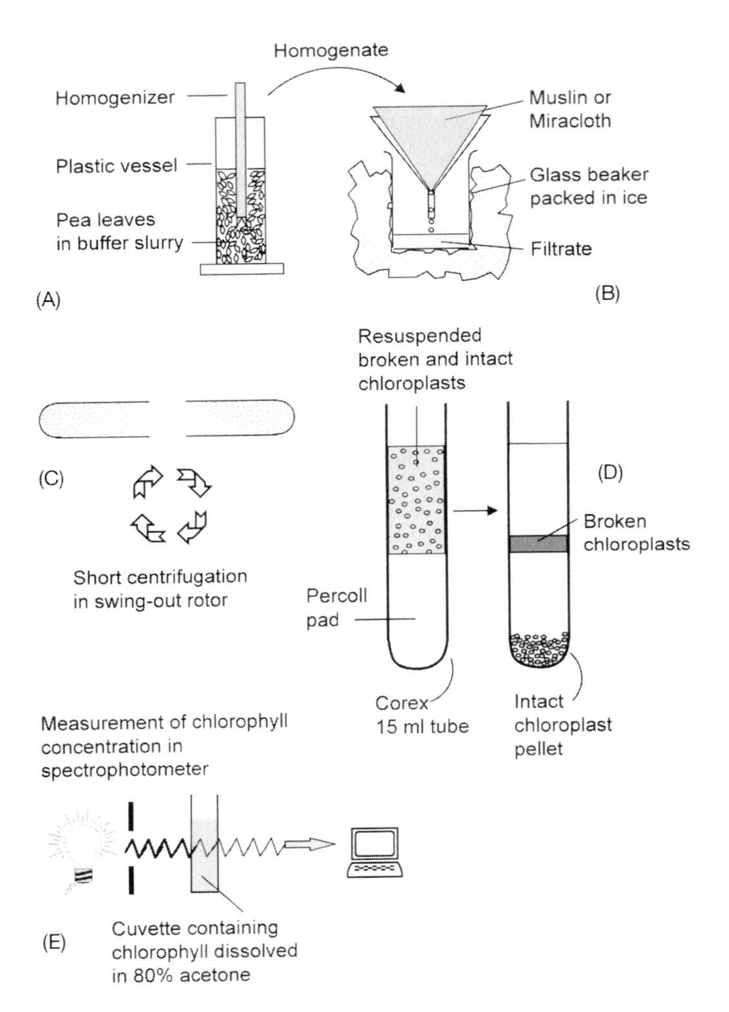


**Figure 2** Summary of the main stages of chloroplast isolation (Protocol 1). (A) Pea seedlings (the unopened leaves, but not the stalks) are homogenized in a cold, Hepes–sorbitol buffer slurry (Protocol 1, step 2). Shown here is a Polytron homogenizer aggregate (Kinematica, type PTA 20 TSM) and a home-made, plastic, cuboid-shaped, homogenization beaker. (B) The homogenate is rapidly filtered through muslin or Miracloth to remove leaf debris (Protocol 1, step 3). The beaker is pre-cooled on ice. (C) The filtered suspension is briefly centrifuged using a swing-out rotor (Protocol 1, step 4). (D) The resuspended chloroplast pellet is loaded on a 35% Percoll pad, and centrifuged using a swing-out rotor, to separate lysed and intact plastids (Protocol 1, step 5). (E) The chlorophyll concentration of the final, washed suspension of chloroplasts is determined by mixing small aliquots of chloroplasts with 80% acetone, and measuring light absorbance in a spectrophotometer (Protocol 1, step 7).

## Protocol 1

# Isolation of intact chloroplasts from pea seedlings

### Equipment and reagents

- 100 ml or 50 ml centrifuge tubes, with rounded rather than pointed bottoms[a]
- 15 ml Corex centrifuge tubes
- Polytron homogenizer (Kinematica AG) or blender of similar efficiency, together with a safe vessel for homogenization
- Muslin or Miracloth (Calbiochem)
- Refrigerated centrifuge with swing-out rotor
- Hepes–sorbitol (HS): 50 mM Hepes–KOH pH 8.0, 330 mM sorbitol[b]
- 35% Percoll (Pharmacia) in HS: 9 ml distilled water, 7 ml Percoll, and 4 ml of 5 × HS, all mixed together is enough for four Percoll pads
- 80% (v/v) acetone in distilled water

### Method

1. Using clean scissors, harvest leaves from pea seedlings (avoiding the lower leaves and stem), and mix with an icy slush of HS, in a ratio of 20 g leaves to 100 ml medium.
2. Homogenize the leaves for the minimum time necessary to produce a suspension of small, evenly-sized leaf fragments, e.g. two 3 sec bursts from a Polytron set at 75% full speed.
3. Strain the homogenate gently through either eight layers of muslin, or two layers of Miracloth, to remove debris. Do not squeeze too hard. If using Miracloth, thoroughly moisten it with a small quantity of HS before filtering the homogenate.
4. Pour the suspension into 50 ml or 100 ml centrifuge tubes. Centrifuge at 4000 g for 1 min in a swing-out rotor. Discard the supernatant in one smooth motion, and while holding the tube upside down, quickly wipe around the inside with a tissue, to remove any froth. At this stage, the pellet is fairly firm, as long as you don't invert the tube more than once.
5. Resuspend each pellet gently in a small volume of HS (4–8 ml), using a cotton swab or a paintbrush (briefly dipped in HS before contacting the pellet), and layer the suspension onto an equal volume of 35% Percoll in HS. This is made easier by tilting the tube (15 ml Corex) at 30° to the vertical, and gently running the chloroplast suspension down the lower side of the tube. Centrifuge at 2500 g for 7–8 min (after the operating speed is attained) in a swing-out rotor, with the brake off. Intact chloroplasts are pelleted, whereas lysed organelles fail to penetrate through the Percoll pad (illustrated in Figures 2D and 3B). The most convenient way to remove the Percoll and green debris is to take a glass Pasteur pipette and place the end just above the layer of lysed chloroplasts—they should come off in a long, billowing string.
6. Wash the pellet in HS to remove residual Percoll (fill the Corex tube up with HS). Centrifuge at 4000 g for 1 min in a swing-out rotor. Resuspend the pellet in a small quantity of HS, e.g. 1 ml.[c]

**Protocol 1** continued

**7** Measure the chlorophyll concentration of the suspension. A rapid estimate may be obtained by adding 5 μl suspension to 1 ml of 80% acetone (do three replicates). After 1 min, spin the tubes for 2 min in a microcentrifuge (top speed), to pellet protein precipitates. Transfer the supernatant to a glass or quartz cuvette, and read the absorbance at 652 nm, against an 80% acetone blank. The chlorophyll concentration (mg/ml) is $A_{652} \times 5.6$ (see ref. 9 for a fuller explanation). Adjust the concentration of the suspension to 1 mg/ml chlorophyll, and use the chloroplasts as soon as possible.

[a] Figure 3A shows an example of a suitable centrifuge tube.

[b] This may be stored as a 5 × concentrated stock (i.e. 250 mM Hepes pH 8.0, 1.65 M sorbitol) in a −20 °C freezer, as Hepes is unstable to autoclaving.

[c] Check the intactness of the organelles by phase contrast microscopy. Intact organelles appear bright green, often with a surrounding halo, whereas broken chloroplasts appear darker and more opaque. The majority of the organelles should be intact, although 50% intactness should give reasonable results.

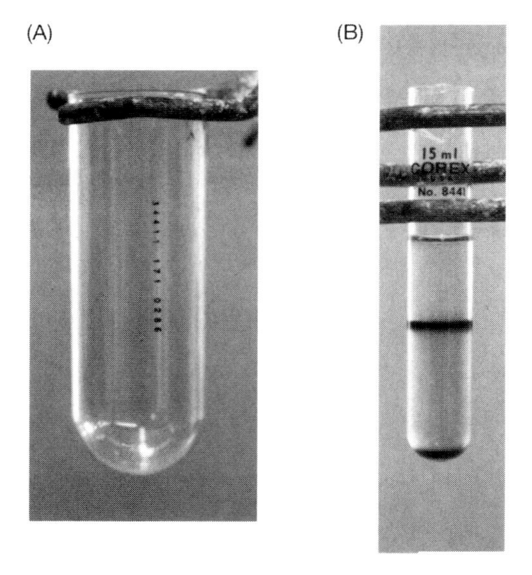


**Figure 3** Centrifuge tubes for isolation of intact chloroplasts. (A) Shown here is an example of a round-bottomed tube, which is handy for the initial centrifugation of the filtered homogenate (Protocol 1, step 4). This one is made from polycarbonate plastic and has a nominal capacity of 100 ml (available from MSE). (B) The photograph shows a 15 ml Corex tube (supported by a clamp) containing a Percoll pad and chloroplasts after centrifugation (Protocol 1, step 5). At the interface between the two layers is a band of lysed chloroplasts, while the pellet at the bottom consists of intact plastids.

## 3 *In vitro* synthesis of nuclear-encoded chloroplast proteins

There are several methods for the generation *in vitro* of synthetic mRNA, following the isolation of a full-length cDNA clone (10). The transcription protocol described in Protocol 2, which can be used with SP6, T7, or T3 RNA polymerase, is particularly handy because the transcription products can be added directly to a cell-free translation system, without first phenol- or salt-extracting the RNA. Try the protocol first without linearizing the vector, but in some cases this may be necessary. If possible, avoid using restriction enzymes which produce 4-base 3′ protruding ends (*Apa*I, *Kpn*I, *Pst*I, *Sac*I, *Sph*I), as these linearized templates produce inhibitory non-coding RNA (11). Where these enzymes must be used, fill in the resulting overhangs with Klenow DNA polymerase.

### Protocol 2

### *In vitro* transcription of cDNA

#### Equipment and reagents

- Nuclease-free plastic-ware
- Water-bath set to 37 °C
- Monomethyl cap ($m^7G(5')ppp(5')G$) (Pharmacia)
- RNA polymerase: SP6 (Gibco) 15 U/μl, T7 (Gibco) 50 U/μl, or T3 (Stratagene) 50 U/μl
- RNase inhibitor (RNasin, Promega)
- Transcription mix containing the following nuclease-free reagents: 40 mM Tris–HCl pH 7.5, 6 mM $MgCl_2$, 2 mM spermidine, 10 mM DTT, 0.5 mM ATP, CTP, and UTP, 50 μM GTP (the NTPs should be pH 7–8), and 100 μg/ml BSA

#### Method

1. Prepare clean DNA in an appropriate transcription vector, either CsCl purified, or from a midi-prep kit (e.g. Qiagen or Promega), at a concentration of 1 μg/μl in water or 10 mM Tris–HCl pH 8.0.[a]
2. Mix at room temperature (cold temperatures can cause DNA to precipitate in the presence of spermidine):
   - 2 μl DNA
   - 15.5 μl transcription mix
   - 20 U RNasin
   - 0.1 U monomethyl cap
   - 1 μl of the appropriate RNA polymerase
3. Incubate for 30 min at 37 °C, add 1 μl of 10 mM GTP, and continue the incubation for a further 30 min. The products of the transcription reaction may be frozen at −70 °C until required.

[a] It is also possible to use mini-prep kits to prepare clean DNA for transcription. Where these involve spin columns, it is essential to make sure that all traces of ethanol wash solution are removed from any internal ledges in the columns, as this can contaminate the final elution buffer, and then inhibit transcription.

Generally the best method for the synthesis of nuclear-encoded chloroplast protein precursors is the wheat germ lysate system, described by Anderson *et al.* (12). The reticulocyte lysate system may also be used, and should be tried when wheat germ lysates fail to translate large mRNA templates. However, several groups have found that reticulocyte lysate can lyse chloroplasts, if used in high concentrations. Bearing this in mind, if using reticulocyte lysate in an intact chloroplast import assay, keep the volume of reticulocyte lysate at or below 10% of the final sample volume. Both wheat germ and reticulocyte lysates are available commercially (e.g. Promega, Amersham).

Follow the manufacturer's instructions when using wheat germ lysate. We find that about 1 μl transcription products (from Protocol 2) per 25 μl translation reaction gives optimum results, with about 30 μCi [$^{35}$S]methionine, or around 25 μCi [$^{3}$H]leucine. [$^{35}$S]cysteine may also be used, but suffers twin drawbacks: namely high expense, and chemical instability, meaning that aliquots should ideally be stored under liquid nitrogen. It is advisable to optimize the translation reaction with respect to the concentration of transcription products (serially dilute the mRNA template first, and add the same volume to each sample); optimal concentrations of potassium and magnesium can also vary drastically with different mRNA species.

## 4 Import of proteins into isolated chloroplasts

Reconstitution of chloroplast protein import is usually straightforward, providing that intact chloroplasts are isolated as described in Protocol 1, and sufficiently labelled precursor proteins can be prepared. Every precursor protein that we have tested so far has been successfully imported into isolated chloroplasts, although with varying efficiencies. Import takes place post-translationally and therefore the precursors are incubated with chloroplasts after translation is complete. Ideally, freshly prepared chloroplasts should be incubated with fresh translation mixtures, but we have found most translation mixtures can still be imported after freezing at −70 °C. However, several rounds of freeze–thawing usually lead to a rapid loss of import-competence, and this is especially relevant for membrane proteins. Another point to consider is the distribution of labelled amino acids in the precursor protein—make sure there is at least one in the mature protein, so the protein may still be detected after import and processing.

The basic import assay conditions, modified from original methods (13) are given in Protocol 3. ATP is required for both binding and import into chloroplasts (14, 15); GTP may also be required by the import machinery (16). It is present in the translation mixture, and will also be generated in the stroma by photophosphorylation when the chloroplasts are illuminated. However we generally add ATP to the assay, and so pre-incubate the chloroplasts at 25 °C for 10 min to allow the stromal ATP concentration to reach the optimal level.

## Protocol 3

# Import of proteins into isolated chloroplasts: basic assay

### Equipment and reagents

- Illuminated water-bath, set at 25 °C
- HS (see Protocol 1)
- 60 mM methionine in 2 × HS[a]
- 60 mM MgATP pH 7.0–8.0 in HS
- Translation mixture
- SB: SDS–PAGE sample buffer

### Method

1. Prepare intact chloroplasts and *in vitro* synthesized proteins as described above.
2. Mix the following components, and pre-incubate for 10 min at 25 °C:
   - 50 μl chloroplasts (equivalent to 50 μg chlorophyll)[b]
   - 20 μl MgATP
   - 55 μl HS
3. Meanwhile, mix and then add the following to the chloroplast suspension after the pre-incubation:
   - 12.5 μl translation mixture
   - 12.5 μl methionine
4. Incubate at 25 °C for 20–60 min in an illuminated water-bath,[c] at an intensity of 300 $\mu E\ m^{-2}\ s^{-1}$. The tubes should be gently shaken or briefly pipetted (cut tips) every 5–10 min to prevent the chloroplasts from settling out.
5. After incubation, dilute the assay with at least 1 ml ice-cold HS, and pellet the chloroplasts by centrifugation at 2000 *g* for 3 min at 4 °C. *Gently* resuspend the chloroplasts in a small volume of ice-cold HS (suggested 120 μl), ready for fractionation. Remove 30 μl (equivalent to one-quarter of the assay), add to SB, and boil for 5 min.[d] This sample contains imported proteins, plus any precursor molecules which are bound to the chloroplast surface.

[a] Unlabelled methionine is included in the import incubation mixture to prevent the high specific activity, labelled methionine in the translation mix from being incorporated into protein by the chloroplast protein synthesis machinery. When using other radiolabelled amino acids, substitute the corresponding 'cold' amino acid for methionine.

[b] The values for the amount of chloroplasts and translation mixture in the assay are intended as a guide; however, too high a chlorophyll concentration may lead to overloading problems during subsequent SDS–PAGE analysis, and the concentration of translation mixture should not exceed 10% of the total assay volume.

[c] An illuminated water-bath may be set up by placing a 150 W light bulb below a glass tank, supported on a transparent Perspex sheet. The bulb should be at least 5 cm away from the plastic, and the air around the bulb must be ventilated with a fan, to prevent the Perspex from melting. The tank water temperature can be regulated by a thermostatted water circulator. If one of these is not available, then monitor the water temperature closely during the incubation, and add cold water as necessary.

[d] Membrane proteins in particular may aggregate during boiling. Therefore, it is wise to test some *in vitro* synthesized precursor in a separate experiment to make sure it will survive boiling. An alternative to boiling is to incubate the sample with SB for 10–20 min at 40 °C, but make sure that any proteases that might have been added to the sample are fully inhibited.

The following method is designed to locate processing intermediates and the mature polypeptide of a thylakoid lumen protein. The strategy is to produce five fractions: total, washed chloroplasts (described above in Protocol 3, step 5), protease-treated chloroplasts, stromal extract, total thylakoids (with some envelope contamination), and protease-treated thylakoids. The latter fraction will demonstrate whether or not the import substrate is protected from protease degradation, within the thylakoid lumen. This method is intended as a guide, which may be adapted to the experimenter's own requirements. An entry into the literature of protein targeting to the chloroplast envelope is provided in the following reviews: Gray and Row (17) and Heins *et al.* (3). Results of some typical chloroplast import assays, some including an inhibitor of transport across the thylakoid membrane, are shown in Figure 4.

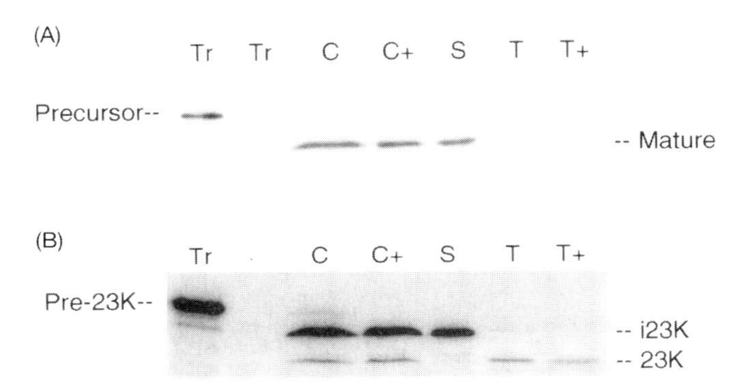


**Figure 4** Import of stromal and thylakoid lumen proteins into intact chloroplasts. (A) A stromal precursor protein was translated in rabbit reticulocyte lysate (track Tr) and incubated with intact pea chloroplasts, according to Protocol 3. After incubation, the chloroplasts were fractionated according to Protocol 4. The incubation resulted in processing of the precursor to a smaller product (the mature protein), which is seen in the total chloroplast track (C), and is protease-protected (track C+). After lysis of the chloroplasts to give stromal and membrane fractions (tracks S and T), it was found that the mature protein fractionated almost entirely with the stroma, a finding which was corroborated by immunological data obtained for the endogenous protein. Track T+ represents protease-treated thylakoids. (B) Pre-23K, the precursor of a thylakoid lumen protein, was translated in wheat germ extract, and incubated with intact pea chloroplasts according to Protocol 3, but in the presence of 2 $\mu$M nigericin (a proton ionophore) and 10 mM KCl. Nigericin dissipates the trans-thylakoidal $\Delta$pH, which strongly inhibits the translocation of proteins, such as 23K, across the thylakoid membrane by the Tat pathway, resulting in the accumulation of stromal intermediates which have not been matured by the action of thylakoidal processing peptidase. Fractionation of the chloroplasts after the incubation revealed the presence of three main radiolabelled bands in the total chloroplasts (track C): the precursor (pre-23K), an intermediate (i23K), and the mature protein (23K). Thermolysin treatment of the chloroplasts (track C+) showed that only i23K and 23K were inside the organelles, and lysis showed that the intermediate was located in the stroma (track S), whereas the majority of the mature protein was protease-protected in the thylakoid lumen (tracks T and T+). If the same incubations were performed in the absence of nigericin, 100% of the radiolabelled protein would be in the mature form, and be located in the thylakoid lumen.

## Protocol 4

# Fractionation of chloroplasts after an import assay

### Reagents

- HS (see Protocol 1)
- Thermolysin (Sigma type X protease): 2 mg/ml in HS; also the same in HM
- 100 mM $CaCl_2$
- HSE: HS containing 50 mM EDTA
- HM: 10 mM Hepes–KOH pH 8.0, 5 mM $MgCl_2$
- HME: HM containing 10 mM EDTA
- 500 mM EDTA pH 8.0
- SB: SDS–PAGE sample buffer

### Method

1. Protease-treat the remaining 90 μl chloroplasts from Protocol 3, step 5 with thermolysin.[a] A suggested method is to add 160 μl thermolysin mixture to the chloroplasts, giving a final volume of 250 μl, and incubate for 40 min on ice. The thermolysin mixture contains 12.5 μl thermolysin at 2 mg/ml, 3.9 mM $CaCl_2$ (2.5 mM final concentration), and HS to 160 μl. It is convenient to make up this mixture for all the samples together, to cut down on individual pipetting actions. The final concentration of thermolysin may be doubled to 0.2 mg/ml if necessary.
2. End the protease digestion by adding 50 μl HSE to chelate the $Ca^{2+}$ ions. Mix well, but gently, and remove 100 μl (one-third of the remaining sample) to a clean tube. Centrifuge this sample at 2000 g for 3 min at 4 °C and resuspend the pellet in 15 μl HSE and 15 μl SB. Boil for 5 min, and then store on dry ice.[b]
3. Centrifuge the remaining 200 μl sample at 2000 g for 3 min at 4 °C, and resuspend the pellet in 60 μl HME using normal uncut tips, to lyse the chloroplasts. The EDTA is present to prevent proteolysis of stromal protein from any remaining thermolysin activity. Incubate the chloroplast lysate for 5 min on ice. Centrifuge at top speed (e.g. 15 000–18 000 g) for 5 min at 4 °C to generate a stromal supernatant and a thylakoid pellet.[c,d]
4. As soon as possible, remove the stromal supernatant, and add it directly to 15 μl boiling SB. Boil for 5 min and then store on dry ice. This stromal fraction is equivalent to *two* aliquots of thylakoid membranes, not one, so remember to load the gel accordingly.
5. Resuspend the thylakoid pellet in 200 μl HM and divide into two equal aliquots. Centrifuge at top speed for 5 min at 4 °C. Resuspend one of the pellets in 15 μl HM and 15 μl SB to give the total thylakoid fraction (boil 5 min, then store on dry ice). Resuspend the other pellet in 100 μl HM, containing 10 μl of 2 mg/ml thermolysin (0.2 mg/ml final concentration) and 2.5 mM $CaCl_2$. Incubate for 40 min on ice.
6. End the protease digestion by the addition of 2 μl of 500 mM EDTA. Centrifuge at top speed for 5 min at 4 °C, and resuspend the pellet in 15 μl HME and 15 μl SB before boiling for 5 min.

**Protocol 4** continued

7 Analyse all the fractions by SDS–PAGE, loading equivalent proportions of each sample. Tricine gels (18) are recommended when the mature protein is between 2 kDa and 16 kDa in size—note that the sample buffer used with Tricine gels differs from that used in standard SDS–PAGE. For analysis by autoradiography, it is suggested that the gels are amplified with a fluorographic reagent (e.g. Amplify from Amersham) before drying them—this can be especially useful for detecting $^3$H-labelled proteins. Gels should be dried on filter paper, rather than between two sheets of plastic, as the plastic will reduce the signal intensity reaching the X-ray film during exposure in a cassette.

[a] Thermolysin is the preferred protease for digesting the outer surface of intact chloroplasts, because it does not alter the envelope permeability (19). However, a control digestion of the precursor protein alone should be carried out in parallel with the assay, to make sure thermolysin will degrade the unimported protein. The protease may be made up in import buffer (HS or HM) at 2 mg/ml and should be stored at −70 °C in single-use aliquots. Calcium is the cofactor required for thermolysin activity, and should only be added to the protease immediately before use.

[b] Supernatants from washes and protease digestions may also be kept for analysis.

[c] After lysis of the chloroplasts, all the membranes are pelleted by centrifugation; hence a small proportion of the total will be envelopes. However, protease treatment of the entire sample in step 1 gets around this potential problem, by digesting unimported precursor proteins attached to the envelope. If the experimenter wishes to separate envelopes from thylakoids post-import and prior to protease digestion, a quick method is described ref. 20.

[d] Make sure each fraction is removed to a clean tube to avoid potential problems with radiolabelled precursor residue, which can smear around the top of the Eppendorf tube during the import assay, and find its way into the sample buffer later, during boiling.

# 5 Import of proteins into isolated thylakoid membranes

## 5.1 The import pathway for thylakoid lumen proteins

Most thylakoid lumen proteins of higher plants and green algae such as *Chlamydomonas reinhardtii* are nuclear encoded, and are synthesized in the cytosol as precursors with an N terminal bipartite presequence. After import into the chloroplast, the first portion of the presequence, termed the envelope transit peptide, is usually, but not always, cleaved off, leaving a signal peptide in tandem with the mature protein domain. All thylakoid signal peptides are superficially very similar in terms of their predicted domain structures, but in fact there are two subsets, which specifically interact with distinct translocases in the thylakoid membrane. After transport across the thylakoid membrane, the signal peptide is removed by a thylakoidal processing peptidase, which is active on the lumenal side of the membrane. The first four thylakoid lumen proteins to be studied were plastocyanin (PC) and the 16, 23, and 33 kDa proteins of the oxygen-evolving complex (16K, 23K, 33K). It turned out that PC and 33K are translocated across the thylakoid membrane by a Sec translocase (2), which is almost certainly inherited from a cyanobacterial progenitor of the chloroplast. On the other hand, 16K and 23K are transported by a ΔpH-dependent translocase, one component of which has been cloned in maize and has recently been shown to have a prokaryotic counterpart (reviewed in refs 21 and 22). This pathway has been renamed 'Tat', which stands for '**t**win-**a**rginine **t**ranslocase', as signal peptides which interact with this translocase contain a twin-arginine motif before their hydrophobic core domains. Nearly all thylakoid lumen proteins studied subsequently have been assigned to one or other of the pathways. To a certain extent, the two pathways may be distinguished by chloroplast import assays, but much more detailed analyses can be carried out by using isolated thylakoids, because conditions such as presence/absence of stromal extract, nucleoside triphosphates (NTPs), etc. are readily manipulated.

## 5.2 The basic import assay

What follows is a simple, light-driven assay for importing precursor proteins into isolated thylakoids, with some suggestions on how to alter conditions to test for the involvement of the Sec or Tat translocases. Two points should be emphasized:

(a) The most efficient *in vitro* import to date has been obtained with isolated pea thylakoids.

(b) For import to take place via the Tat pathway, the sole energy requirement is the thylakoidal ΔpH. ATP or other NTPs are not required, nor is stromal extract (the envelope transit peptide does not have to be removed from pre-16K and pre-23K prior to import into isolated thylakoids). Transport via the Sec pathway requires stromal SecA, and ATP as well as the thylakoid-associated Sec apparatus. Additional soluble factors for some Sec substrates cannot be ruled out at this stage. The thylakoidal ΔpH is not a prerequisite for transport by the Sec pathway, but may stimulate the transport of some proteins under certain *in vitro* conditions. Thus, for any new thylakoid lumen protein, it is recommended to try import in the presence and absence of stromal extract first of all.

## Protocol 5

### Import of proteins into isolated thylakoids

#### Equipment and reagents

- Illuminated water-bath set to 25 °C
- HM
- 2 mg/ml thermolysin in HM
- 100 mM $CaCl_2$
- 500 mM EDTA pH 8.0
- Translation mixture
- SB: SDS–PAGE sample buffer

**Protocol 5** continued

### Method

1 Prepare a pellet of intact pea chloroplasts as described in Protocol 1, steps 1–7.

2 Lyse the chloroplasts by resuspending the pellet in ice-cold HM at a chlorophyll concentration of 1 mg/ml.[a] Leave on ice for 5 min to ensure complete lysis.

3 Centrifuge at top speed (e.g. 15 000–18 000 g) for 5 min at 4 °C to generate a stromal supernatant and a thylakoid pellet. Keep the stromal extract on ice until required.

4 Wash the thylakoids twice in HM, taking care not to blow bubbles in the viscous suspension. As a starting point, allow 20 μg chlorophyll per assay: for ease of handling, wash only as many membranes as are needed for the experiment, plus a little extra in case of losses. Resuspend the final pellet in either HM or stromal extract, to a concentration of 0.5 mg/ml chlorophyll.

5 Set up the import incubation:
- 40 μl thylakoid suspension
- 5 μl translation mixture
- 5 μl HM

6 Incubate at 25 °C for 20–30 min under illumination (300 μE $m^{-2}$ $s^{-1}$).

7 After incubation, the thylakoids should be washed. However, the experimenter may be interested to look at a sample of the unwashed membranes. If this is the case, remove one-third of the incubation **before** the wash step, and add it direct to an equal volume of SB, boil for 5 min, and store on dry ice. Otherwise, add 1 ml ice-cold HM to the incubation, centrifuge at top speed for 5 min at 4 °C, and resuspend the pellet in a small volume (e.g. 40 μl) HM.

8 Divide the suspension in half (each to clean tubes). To one, add an equal volume of SB, boil for 5 min, and store on dry ice. To the other, add thermolysin to a final concentration of 0.2 mg/ml and $CaCl_2$ at 2.5 mM—a suitable volume for the protease digestion is 100 μl. Incubate for 40 min on ice, and halt the digestion by the addition of 10 mM EDTA. Pellet the membranes at top speed for 5 min at 4 °C, then resuspend the pellet in HM and 10 mM EDTA before quickly adding boiling SB (to ensure rapid inactivation of the protease).

[a] The amount of stromal extract added to the assay may be increased by lysing a more concentrated suspension of chloroplasts, e.g. at 2 mg/ml chlorophyll.

## 5.3 Some variations on the basic assay

Nucleoside triphosphates (largely present in the translation mixture) may be removed by the enzyme apyrase (Sigma, grade VI); a recent example of the method is described in ref. 23. Alternatively, stromal extract and translation mixture can be gel filtered on small, disposable Sephadex columns (obtained from Pharmacia), which then enables the researcher to add back specific NTPs, analogues, etc. Examples of this technique are described in refs 24 and 25.

The thylakoidal ΔpH may be dissipated by 2 μM nigericin (26). Note that this proton ionophore requires $K^+$ ions to function, but if the HM import buffer pH has previously been adjusted with KOH, there should be sufficient $K^+$ already present. Otherwise, add 10 mM KCl to the assay.

The translocation factor SecA can be inhibited by 10 mM sodium azide, especially in the presence of low concentrations of ATP (27). However, sensitivity to azide, or a lack of it, is not enough on its own to assign a protein to a particular translocation pathway, especially in the light of results reported by Leheny *et al.* (28) who have shown that 16K translocation is inhibited by azide, despite the general acceptance that this protein is a substrate for the ΔpH-dependent (Tat) pathway.

## 5.4 Analysis of nuclear-encoded thylakoid membrane proteins

The basic methods described in this chapter so far have been framed as though the investigator is importing a thylakoid lumen protein. This is simply because it is quite straightforward to show that a lumen protein has been correctly targeted—if fully translocated across the membrane, then the protein will be completely protected from protease degradation. However, Protocols 3, 4, and 5 may be easily adapted for studying the insertion of thylakoid membrane proteins. To date, most thylakoid membrane proteins may be divided into two main groups:

(a) Those which are synthesized with an N terminal envelope transit peptide, which serves to target the protein into the chloroplast. Targeting to the thylakoid membrane is accomplished by means of information contained within the mature protein. The classic example of this group is Lhcb1 which is the major chlorophyll *a*/*b* binding protein of light harvesting complex II.

(b) Those which are synthesized with an N terminal **bipartite** presequence, closely resembling that of thylakoid lumen proteins. It is currently thought that the signal peptide assists in the insertion of the mature protein, but not through interactions with either the Sec or Tat translocases. An example of this group is PsbW, which is a small subunit of photosystem II.

The most intensively studied thylakoid membrane protein is Lhcb1, which requires the chloroplast signal recognition particle plus its receptor, GTP hydrolysis, and membrane-bound components for correct insertion (29, 30). Lhcb1 can be efficiently imported into intact pea chloroplasts, and also inserted into isolated thylakoid membranes, but strictly in the presence of stromal extract and NTPs. However, not all precursors that are structurally similar to Lhcb1 have identical insertion requirements (31), and the unravelling of the various insertion mechanisms of the huge number of thylakoid membrane proteins is a challenging and fascinating problem. Proteins from group (b) do not need NTPs, a ΔpH, or any soluble factors for their insertion, and will even insert efficiently into thylakoid membranes which have been pre-digested with trypsin. Regardless of the insertion mechanism of any protein to be investigated though, there are two important points to be considered before embarking on any experimental work:

(a) Membrane proteins are hydrophobic, and thylakoid membranes are quite 'sticky'. This means that a precursor can appear to be inserted into isolated thylakoids when in reality it has either stuck non-specifically to the membrane surface, or has aggregated and then sedimented with the membranes during centrifugation.

(b) It is necessary not only to check that a protein has inserted into the membrane, but that it has done so with its correct topology.

A reasonable strategy is to import the precursor of interest into intact chloroplasts first of all. The advantage of this *in organello* approach is the high probability that all the protein fractionating with the thylakoids (Protocol 4, step 5) will be correctly inserted. The next problem is to find a protease (and concentration) that will completely digest the precursor protein (a sample of the translation mixture), but will give a defined and reproducible degradation pattern for the integrated protein and does not chew through to the thylakoid lumen. A useful review of proteases that can be used to determine the topology of membrane proteins can be found in ref. 32. In our experience, trypsin, proteinase K, and thermolysin are good proteases to test in the first instance. **Full inhibition of these proteases at the end of the incubation is of prime importance**, since addition of SDS-PAGE sample buffer will expose those parts of the protein which were previously membrane protected, and many proteases work happily in the presence of SDS. Of course, this method relies on the assumption that the radiolabelled protein has been inserted exactly as it would be *in planta*. In practice, there is often some corroborating information available in the literature to help decide if the imported protein really does behave like its endogenous counterpart. When a satisfactory set of protease digestion conditions has been set up, it should then be possible to proceed to looking at insertion of the precursor into isolated thylakoid membranes, the aim being to observe similar degradation products from the putative integrated protein. Figure 5 shows the results of an assay for the insertion of PsbY into isolated thylakoids. The degradation products from thermolysin digestion exactly match those seen from protein imported into intact chloroplasts, and also agree with theoretical predictions for the transmembrane topology.

Chemical extraction techniques are another tool to help the researcher decide whether a protein is associated with the membrane or not. Breyton *et al.* (33) describe a range of different treatments, which they applied to *Chlamydomonas reinhardtii* thylakoid membranes. Of these, we have found urea extraction to be particularly useful in its severity; if carried out carefully, any protein detected in the membrane pellet (and therefore resistant to extraction) is probably inserted in the membrane (Figure 6). Protocol 6 is adapted from ref. 33, and may be used to extract thylakoid membranes obtained from either a chloroplast import assay (Protocol 4, step 5), or from an isolated thylakoid insertion assay.

(A)

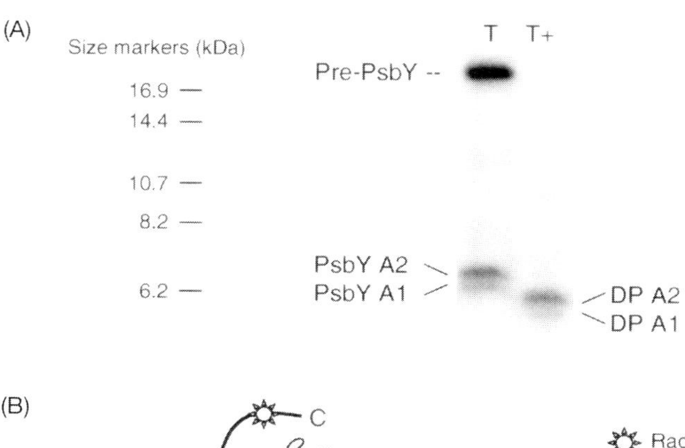


(B)

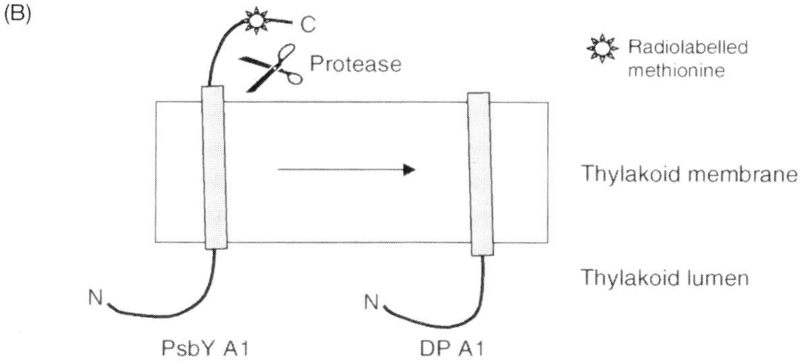


**Figure 5** Insertion of a membrane protein into isolated thylakoids. (A) *Arabidopsis thaliana* pre-PsbY is a nuclear-encoded polyprotein, which is processed by thylakoids to yield two much smaller polypeptides (PsbY A1 and PsbY A2) each of which has one transmembrane span, oriented with the N terminus in the thylakoid lumen. In this experiment, 5 μl pre-PsbY, translated *in vitro* by a wheat germ extract and labelled with [$^{3}$H]leucine, were mixed with washed thylakoid membranes and incubated in the light, for 20 min at 26 °C. After the incubation, the membranes were washed once with 10 mM Hepes–KOH pH 8.0, 5 mM $MgCl_2$ (HM) and divided into two samples. The first sample was mixed with sample buffer and analysed directly (track T), and the second sample was digested with 100 μl of 0.2 mg/ml thermolysin, 2.5 mM $CaCl_2$ all dissolved in HM, for 40 min on ice (track T+). This thermolysin digestion was halted by the addition of 10 mM EDTA, the membranes washed once, and then resuspended in HM, 10 mM EDTA, and sample buffer. A Tricine gel system (16) was used to separate the proteins, which were then visualized by fluorography. In the protease treatment track the mature polypeptides PsbY A2 and A1 have each been clipped to smaller degradation products, DP A2 and DP A1, while the non-inserted precursor protein has been completely degraded. (B) A schematic representation of the protease treatment described in (A). The C terminus of inserted PsbY A1 is accessible to protease, leaving a protected fragment, DP A1, in the thylakoid membrane. A theoretical prediction that the C terminus is exposed on the stromal side of the thylakoid membrane was confirmed by a comparison of [$^{35}$S]methionine and [$^{3}$H]leucine labelled PsbY. PsbY A2 and A1 each contain a single methionine residue near the C terminus of the protein, but several evenly distributed leucine residues. When inserted [$^{35}$S]methionine labelled PsbY is digested with thermolysin, the polypeptides almost completely disappear from the gel, whereas two distinct degradation products result from digestion of $^{3}$H-labelled PsbY.

## Protocol 6

# Urea extraction of thylakoid membranes

### Equipment and reagents

- Bench-top ultracentrifuge, such as a Beckman TL100, with a TLA100.3 fixed angle rotor
- Urea-resistant centrifuge tubes, e.g. Beckman Polyallomer 1.5 ml Eppendorfs[a]
- TN: 20 mM Tricine–NaOH pH 8.0
- Thylakoid membranes (10–20 μg chlorophyll per sample) from either a chloroplast import assay (Protocol 4, step 5), or an isolated thylakoid import/insertion assay (Protocol 5, step 7), stored on ice
- UT: 6.8 M urea/20 mM Tricine–NaOH; freshly made
- SB: SDS–PAGE sample buffer

### Method

1. Transfer the thylakoids to clean Eppendorf tubes. Wash the thylakoids in ice-cold TN. Recover the membranes by centrifugation at top speed in a cold microcentrifuge (e.g. 15 000–18 000 *g* for 5 min). Remove every last drop of supernatant.
2. Resuspend the pellets in 100 μl UT and incubate for 10 min at room temperature.
3. Freeze the samples, then defrost at room temperature. Repeat this step.
4. Place the tubes in appropriate rotor adaptors and centrifuge at 120 000 *g* for 15 min at 4 °C.
5. Immediately after centrifugation, carefully remove the tubes and withdraw the top 80 μl supernatant, avoiding the fairly loose pellet. Keep this as the 'urea supernatant' sample (adding SB to it), which contains the extracted material. Gently remove the remaining 20 μl supernatant without drawing up any of the pellet. This 20 μl can be discarded. Remember that the 80 μl supernatant is not the full, original volume when you come to load the samples on the gel.
6. Resuspend the pellet in 100 μl UT and repeat steps 2–5. Usually, very little material is extracted in this second round.
7. Resuspend the final pellet in TN, and add SB. This is the membrane pellet, containing urea-resistant protein.[b]

[a] Normal Eppendorf tubes start to fail if centrifuged at 120 000 *g* in the presence of 6.8 M urea.

[b] There is a risk that aggregated precursor protein from the translation mixture could sediment with the membranes during the centrifugation at 120 000 *g* (step 4), giving the appearance that it is resistant to urea extraction from the membrane. We have noticed this occasionally. There is a precautionary step, which can be taken right at the beginning of the experiment, before setting up the isolated thylakoid insertion assay. After translation is completed, centrifuge the translation mixture (e.g. 120 000 *g* for 10 min at 4 °C) and use the supernatant so obtained in the thylakoid insertion assay. The precautions taken in step 1 (transferring to a fresh tube, and washing the thylakoids with TN) should usually be adequate though.

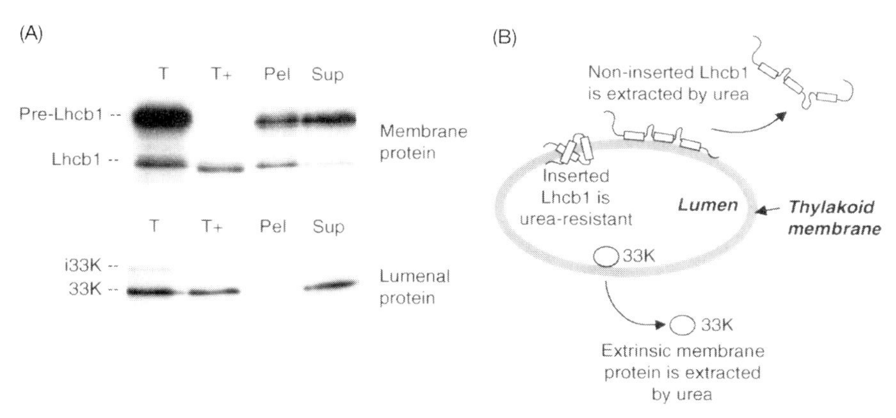


**Figure 6** Use of extraction techniques to test whether proteins are inserted into membranes. (A) Washed thylakoids suspended in stromal extract were incubated with 5 μl *in vitro* translated petunia pre-Lhcb1 (a triple-spanning membrane protein) or wheat i33K (intermediate-sized 33K, a thylakoid lumen protein together with its N terminal signal peptide, but lacking the envelope transit signal). After the incubation, the membranes were washed with 10 mM Hepes–KOH pH 8.0, 5 mM $MgCl_2$ (HM), and then either analysed directly (tracks T), after digestion with 0.2 mg/ml thermolysin, 2.5 mM $CaCl_2$, all in 100 μl HM (tracks T+), or after extraction with 100 μl of 6.8 M urea, 20 mM Tricine–NaOH (tracks Pel). For both proteins, incubation with thylakoids results in processing to the mature size, as seen in tracks T. Protease treatment of the membranes gives a slightly smaller degradation product for Lhcb1, indicating that a part of the protein is exposed on the surface of the membrane. The 33K mature protein is fully protease protected, which means it has been transported across the thylakoid membrane and is in the lumen. After urea extraction, none of the 33K is present in the membrane pellet (track Pel); all of it has been extracted into the supernatant (track Sup). On the other hand, a proportion of the pre-Lhcb1, and the vast majority of the mature Lhcb1 fractionates with the membrane pellet, and is therefore resistant to urea extraction. The extractable pre-Lhcb1 may be, for example, protein that is stuck to the surface of the thylakoid membrane. The urea-resistant **pre**-Lhcb1 is likely to be inserted in the membrane, but still awaiting processing by stromal processing peptidase (SPP)—some membrane proteins are able to be cleaved by SPP after being inserted in the thylakoid membrane. (B) Summary of the urea extraction data represented in (A). Proteins firmly anchored in the membrane are resistant to urea extraction, whereas extrinsic, or lumenal proteins are washed from the thylakoids. Our data suggest that very small, single-spanning membrane proteins, or protease degradation products are less resistant to urea extraction compared to larger, or multi-spanning membrane proteins, but still give qualitatively different results from non-membrane proteins.

Chemical extraction techniques may also be combined with protease treatments. For example, after urea extraction, the thylakoids may be washed with HM buffer and resuspended in a solution of protease, which can give a cleaner result than proteolysis alone. We hope the above discussions and protocols will give the reader some helpful ideas and act as a basic guide to investigating protein targeting in chloroplasts.

## References

1. Cline, K. and Henry, R. (1996). *Annu. Rev. Cell Dev. Biol.*, **12**, 1.
2. Robinson, C. and Mant, A. (1997). *Trends Plant Sci.*, **2**, 431.

3. Heins, L., Collinson, I., and Soll, J. (1998). *Trends Plant Sci.*, **3**, 56.
4. Chen, L.-J. and Li, H. (1998). *Plant J.*, **16**, 33.
5. Rensink, W. A., Pilon, M., and Weisbeek, P. (1998). *Plant Physiol.*, **118**, 691.
6. Dahlin, C. and Cline, K. (1991). *Plant Cell*, **3**, 1131.
7. Sutton, A., Sieburth, L. E., and Bennett, J. (1987). *Eur. J. Biochem.*, **164**, 571.
8. Nielsen, V. S., Mant, A., Knoetzel, J., Møller, B. L., and Robinson, C. (1994). *J. Biol. Chem.*, **269**, 3762.
9. Hipkins, M. F. and Baker, N. R. (1986). In *Photosynthesis energy transduction: a practical approach* (ed. M. F. Hipkins and N. R. Baker), p. 51. IRL Press, Oxford.
10. Melton, D. A., Krieg, P., Rabagliciti, M. R., Maniatis, T., Zinn, K., and Green, M. R. (1984). *Nucleic Acids Res.*, **12**, 7035.
11. Schenborn, E. T. and Mierendorf, R. C. (1985). *Nucleic Acids Res.*, **13**, 6223.
12. Anderson, C. W., Straus, J. W., and Dudock, B. S. (1983). In *Methods in enzymology* (ed. R. Wu, L. Grossman, and K. Moldave), Vol. 101, p. 635. Academic Press, London.
13. Grossman, A. R., Bartlett, S. G., Schmidt, G. W., Mullett, J. E., and Chua, N.-H. (1982). *J. Biol. Chem.*, **257**, 1558.
14. Olsen, L. J., Theg, S. M., Selman, B. R., and Keegstra, K. (1989). *J. Biol. Chem.*, **264**, 6724.
15. Theg, S. M., Bauerle, C. B., Olsen, L. J., Selman, B. R., and Keegstra, K. (1989). *J. Biol. Chem.*, **264**, 6730.
16. Kessler, F., Blobel, G., Patel, H. A., and Schnell, D. J. (1994), *Science*, **266**, 1035.
17. Gray, J. C. and Row, P. E. (1995). *Trends Cell Biol.*, **5**, 243.
18. Schägger, H. and von Jagow, G. (1987). *Anal. Biochem.*, **166**, 368.
19. Cline, K., Werner-Washburne, M., Andrews, J., and Keegstra, K. (1984). *Plant Physiol.*, **75**, 675.
20. Mant, A. and Robinson, C. (1998). *FEBS Lett.*, **423**, 183.
21. Settles, A. M. and Martienssen, R. (1998). *Trends Cell Biol.*, **8**, 494.
22. Dalbey, R. E. and Robinson, C. (1999). *Trends Biochem. Sci.*, **24**, 17.
23. Thompson, S. J., Robinson, C., and Mant, A. (1999). *J. Biol. Chem.*, **274**, 4059.
24. Mant, A., Schmidt, I., Herrmann, R. G., Robinson, C., and Klösgen, R. B. (1995). *J. Biol. Chem.*, **276**, 23275.
25. Hoffman, N. E. and Franklin, A. E. (1994). *Plant Physiol.*, **105**, 295.
26. Mills, J. D. (1986). In *Photosynthesis energy transduction: a practical approach* (ed. M. F. Hipkins and N. R. Baker), p. 143. IRL Press, Oxford.
27. Knott, T. G. and Robinson, C. (1994). *J. Biol. Chem.*, **269**, 7843.
28. Leheny, E. A., Teter, S. A., and Theg, S. M. (1998). *Plant Physiol.*, **116**, 805.
29. Tu, C.-J., Schuenemann, D., and Hoffman, N. E. (1999). *J. Biol. Chem.*, **274**, 27219.
30. Moore, M., Harrison, M. S., Peterson, E. C., and Henry, R. (2000). *J. Biol. Chem.*, **275**, 1529.
31. Kim, S. J., Jansson, S., Hoffman, N. E., Robinson, C., and Mant, A. (1999). *J. Biol. Chem.*, **274**, 4715.
32. Pratt, J. M. (1989). In *Proteolytic enzymes: a practical approach* (ed. R. J. Beynon and J. S. Bond), p. 181. IRL Press, Oxford.
33. Breyton, C., de Vitry, C., and Popot, J.-L. (1994). *J. Biol. Chem.*, **269**, 7597.

# Chapter 6
# Transport into and out of the nucleus

Patrizia Fanara, Adam C. Berger,
Deanna M. Green, Henry Hagan,
Michelle T. Harreman, Kavita A. Marfatia,
B. Booth Quimby, Maureen A. Powers, and
Anita H. Corbett
Departments of Biochemistry and Cell Biology, Emory University School of Medicine, 1510 Clifton Road, NE Atlanta, GA 30322, USA.

## 1 Introduction

How macromolecules are transported between the cell nucleus and the cytoplasm is a general question of great importance to cell biology. Diverse macromolecules, including proteins, mRNAs, tRNAs, and ribosomal subunits, for example, are specifically transported between the cytoplasm and the nucleus and many different techniques have been developed to study these processes. This chapter focuses on those techniques that are specific to the transport process rather than the more general methods that have been used to understand the molecular mechanisms that underlie nucleocytoplasmic trafficking. We present:

(a) Techniques that have been developed to look at the intracellular localization of macromolecules as well as specific assays that have been developed to examine dynamic macromolecular trafficking.

(b) Methods that have been utilized to examine the role of the small GTP binding protein Ran and to provide a general framework for the interactions that occur between soluble nuclear transport factors and components of the nuclear pore complex.

The goal is for a researcher to be able to examine various aspects of nuclear transport using the assays described. This should be possible through the application of well tried and successful protocols presented in this chapter. In general, all the described *in vivo* assays have been used to study the intracellular localization of macromolecules (Protocols 1–7), whereas *in vitro* assays have been mainly used to characterize the molecular interactions that comprise the transport pathways (Protocols 8–10). We also reference a number of other approaches that have been extremely useful in studying nuclear transport in the past as well as novel methods that are emerging as powerful tools for future studies.

# 2 Nuclear transport assays

## 2.1 *In vitro* protein import in permeabilized cells

Analysis of protein import in permeabilized tissue culture cells is the primary tool that has been used to identify proteins that are important for nuclear transport (1). It is also an assay of general use for those who want to determine how their substrate of interest is transported into the cell nucleus. In this assay the cell membrane is permeabilized using digitonin and the cytoplasm washed away without damage to the nucleus.

### 2.1.1 Preparation of NLS substrate

In Protocol 1, the preparation of a classical nuclear localization signal (NLS) containing substrate is described. This protocol describes the preparation of a fluorescently labelled protein conjugate, which serves as a substrate for the *in vitro* nuclear transport system. First, a carrier protein is labelled with rhodamine. Then a peptide encoding a nuclear localization signal is covalently coupled to the fluorescent carrier. This substrate serves as a positive control for the *in vitro* import assay. Protocol 2 describes the *in vitro* import assay in detail where the substrate (or another substrate of interest) is incubated with the permeabilized cells either in the presence of cytosol or with purified proteins.

## Protocol 1

## Preparation of nuclear import substrate

### Equipment and reagents

- Two Biogel P6-DG, 7–8 ml bed volume columns, or other equivalent desalting gel (Sephadex G25, etc.), or pre-packed column
- Human serum albumin (HSA) (Calbiochem)
- Tetramethylrhodamine isothiocyanate (TRITC, isomer R) (Sigma)
- 0.1 M sodium bicarbonate pH 9.0
- 0.1 M sodium phosphate pH 6.0
- Dimethylformamide (DMF)
- Dimethyl sulfoxide (DMSO)
- SV40 T antigen NLS peptide: CTPPKKKRKV (or NLS peptide of choice)
- *m*-Maleimidobenzoyl-*N*-hydroxysuccinimide (MBS) (Pierce)
- Substrate buffer: 20 mM Hepes pH 7.5, 50 mM KCl
- Bio-Rad protein assay (if needed)

### Method

1 Dissolve 10 mg HSA in 1 ml of 0.1 M sodium bicarbonate pH 9.0. Just before use, dissolve TRITC in DMSO at 1 mg/ml. Add 100 μl of TRITC solution to the 10 mg/ml HSA and incubate at room temperature for 1 h, protected from light.[a]

**Protocol 1** continued

2 Dissolve MBS at 10 mg/ml in DMF. Add 250 μl of MBS to the TRITC–HSA and incubate at room temperature for a further 1 h, protected from light.[b]

3 While the reaction is incubating, equilibrate the first Biogel column in 0.1 M sodium phosphate pH 6.0.

4 Load the reaction onto the equilibrated column, collect the 1 ml flow-through, and discard. Elute the column with 0.1 M sodium phosphate pH 6.0, collecting 0.4 ml fractions. Pool the fractions from the first pink peak to give a total volume of 1.5–2 ml. These will contain the rhodamine labelled HSA–MBS. The second pink peak will contain free TRITC and need not be eluted from the column. If necessary, the protein peak can be located by adding 10 μl of each fraction to 0.5 ml of Bio-Rad protein assay reagent (1:5 with water). Colour will appear almost immediately. Pool the fractions that yield the blue protein peak.

5 The HSA concentration should be approx. 5 mg/ml in 0.1 sodium phosphate pH 6.0. This can be confirmed by protein assay.

6 Save a 5 μg aliquot of TRITC–HSA–MBS for later gel analysis.

7 To conjugate the NLS peptide, dissolve NLS peptide at 10 mg/ml in 0.1 M sodium phosphate pH 6.0. It is generally easiest to make up 0.5–1.0 ml of solution, the excess can be stored at −80 °C for later use. Solid peptides may contain some residual acid; adjust the pH of the peptide solution to 6.0 by addition of 2 μl of 1 N NaOH at a time, monitoring the pH by spotting 5–10 μl on a pH strip.[c]

8 Add 250 μl (2.5 mg) of peptide solution to 1 ml of TRITC–HSA–MBS one drop at a time with mixing. The solution may become turbid but this will not prevent coupling. Incubate for 1–1.5 h at room temperature with mixing and protected from light.[d]

9 While the reaction is incubating, equilibrate the second Biogel column in substrate buffer.

10 Load the reaction onto the equilibrated column to remove free peptide. Collect 1.5 ml of flow-through, then elute the column using substrate buffer collecting 0.4 ml fractions. Again, the peak of pink fractions is pooled, or the peak can be identified by protein assay as above. The final protein concentration of TRITC–HSA–NLS should be approx. 1 mg/ml. If necessary, the concentration can be brought up to 1 mg/ml with any centrifugal concentrator.[e]

11 Freeze small aliquots of transport substrate in liquid $N_2$ and store at −80 °C. Avoid repeated freeze-thawing. After thawing, aliquots are stable for a month stored at 4 °C and protected from light.

[a] This reaction couples TRITC to the epsilon amino group of lysines on HSA.

[b] At this pH, the primary reaction is the *N*-hydroxysuccinimide (NHS) group of MBS with the epsilon amino group of lysines. While some lysines have already been modified by TRITC, there are still a sufficient number available for MBS addition. An alternative to MBS is the use of sulfo-MBS (Pierce) which is water soluble and avoids the use of DMF.

[c] The SV40 NLS peptide usually dissolves readily under these conditions. Some other peptides are not fully soluble even when adjusted to pH 6.0, however coupling will still occur.

**Protocol 1** continued

[d] At this pH, the maleimide group of MBS reacts with the cysteine at the N terminus of the NLS peptide. Since the MBS is linked to HSA, this results in crosslinking of the peptide to the protein.

[e] The extent of coupling is determined by electrophoresis through a 7.5 or 8% acrylamide gel. The gel can be stained with Coomassie brilliant blue to visualize protein or the fluorescent substrate can often be photographed directly on a UV light box. The average molecular weight of the transport substrate is compared to the average molecular weight of TRITC–HSA–MBS from step 5. This is necessary because the addition of MBS causes a significant change in molecular weight of the HSA as well as broadening of the band due to variability in the amount of MBS/protein. The difference in average molecular weight is then divided by the weight of the peptide to give the average number of peptides per HSA. Typically coupling is approx. 12 peptides per HSA.

### 2.1.2 *In vitro* protein import assay

Protocol 2 describes an assay for import into the nuclei of tissue culture cells (Protocol 2B) following selective permeabilization of the plasma membrane (Protocol 2A). This assay can be used with a complete exogenous cytosol, as described below, to test the import competence of a particular substrate. Alternatively, the assay can be used with recombinant import factors to determine the requirement for specific factor(s) in the transport of a substrate of interest. Many sources of cytosol have been used with this assay including the membrane-free fraction of *Xenopus* egg extract, rabbit reticulocyte lysate, and HeLa cell cytosol. Protocols for preparation of each of these extracts are more extensive than can be covered here. Detailed descriptions of the preparation of *Xenopus* extract (2) and HeLa cytosol (1) are available. Rabbit reticulocyte lysate is commercially available (Promega and others).

## Protocol 2

## Nuclear protein import assay in permeabilized cells

### Equipment and reagents

- Fluorescent microscope with Plan-apo oil immersion ×60 objective equipped with 510–560 nm excitation filter and 573–648 nm emission filter for viewing TRITC labelled substrate, and 360 nm excitation filter and 460 nm emission filter for viewing Hoescht
- Transport buffer: 20 mM Hepes pH 7.4, 110 mM potassium acetate, 5 mM sodium acetate, 2 mM magnesium acetate, 1 × complete protease inhibitors (Roche)
- Cytosol: HeLa cell cytosol, rabbit reticulocyte, *Xenopus* egg extract
- Paraformaldehyde (Polysciences)
- Digitonin (Calbiochem): 40 mg/ml stock in DMSO, stored at −20 °C
- Hoechst (bisbenzimide; Calbiochem): 10 mg/ml stock, stored at 4 °C
- Cultured adherent cells (HeLa, NRK, BRL, etc.) grown on coverslips
- 10 × ATP regeneration system: 200 mM phosphocreatine (Sigma), 20 mM ATP, 50 μg/ml creatine phosphokinase in 50% glycerol (Sigma); individual stocks stored at −20 °C

**Protocol 2** continued

## A. Permeabilization and preparation of cells

1 Following trypsinization, grow cells on 12 mm coverslips in the appropriate medium for at least 24 h before use to ensure good adherence. Plate cells at a density such that they will be subconfluent at the time of assay.[a]

2 At the time of assay, remove medium and rinse coverslips once with ice-cold transport buffer. Remove buffer and add 1 ml per well of transport buffer containing 40 μg/ml digitonin (diluted from 40 mg/ml stock). Incubate cells in this permeabilization solution on ice for 5 min. After this incubation, rinse cells once with cold transport buffer to remove the digitonin, add 1 ml per well of cold transport buffer, and incubate cells on ice for 15 min to wash out the cytosol.[b,c]

## B. Import assay

1 Prepare assay mixes on ice. A typical 50 μl reaction mix contains: 5 mg/ml *Xenopus* egg extract or 2–3 mg/ml HeLa cytosol, 1 × ATP regeneration system (20 mM phosphocreatine, 2 mM ATP, 5 μg/ml creatine phosphokinase), 1–2 μg fluorescent transport substrate, any desired additions such as inhibitors, competitors, GTP analogues, etc., and transport buffer to bring the total volume to 50 μl. Before initiation of the assay, place reaction mixes as drops on Parafilm in a humidified chamber (a plastic Petri dish lined with dampened filter paper) and incubate at room temperature for 5 min to come to temperature.[d]

2 Recover treated and washed coverslips from wells using forceps and blot the edge on filter paper to remove excess buffer. Invert coverslips onto a 50 μl drop of assay mix. Incubate reactions in the dark at room temperature for 20 min.

3 At the completion of the import reaction, blot coverslips to remove excess reaction mix and rinse twice in ice-cold transport buffer. After blotting to remove excess buffer, invert coverslips onto a 50 μl drop of 4% paraformaldehyde in PBS and incubate for 10 min at room temperature to fix. Blot coverslips again to remove the fixative and wash twice for 5 min in PBS. The first wash contains the DNA dye, Hoechst, at 1 μg/ml concentration to enable later visualization of the nuclei in the UV channel of the fluorescent microscope.

4 Blot coverslips to remove excess liquid and mount on a drop of mounting medium.[e]

[a] Pre-treatment of the coverslips is dependent upon the particular cell line used. We have found that HeLa and NRK cell lines adhere sufficiently well to ethanol washed glass coverslips; for these cells it is not necessary to coat the glass with poly-lysine or other substrates. However, the appropriate culture conditions must be assessed for each specific cell type.

[b] The detergent digitonin is specific for cholesterol and thus will selectively permeabilize the plasma membrane, which has the highest cholesterol content in the cell. However, different cell types can vary in their sensitivity to digitonin, so the concentration and time of treatment should be tested empirically for any new cell type used in the assay.

[c] It is important to establish the integrity of the nuclear envelope following digitonin permeabilization, especially so when a new cell type is used. To do this, immunofluorescence is performed on duplicate coverslips using a nuclear antigen-specific antibody. One convenient

**Protocol 2** continued

and commercially available reagent is anti-RCC1, the exchange factor for the Ran GTPase (Santa Cruz Biotechnologies; Cat. No. sc1161 or sc1162). Following digitonin permeabilization, duplicate coverslips are fixed and then one of the pair is permeabilized in the presence of 0.2% Triton. The cells are incubated with anti-RCC1, followed by a fluorescent secondary antibody, in the absence of detergent. If the nuclei of the permeabilized cells remained intact following digitonin treatment, the signal corresponding to the nuclear RCC1 protein will be observed only in those samples where Triton was added.

[d] If *Xenopus* egg extract is used, this must be diluted to a protein concentration of 20 mg/ml or less in the reaction mix. Higher concentrations of this extract may cause cells to detach from the coverslip. The amount of transport substrate is approximate; this should be optimized for each batch of substrate. Although not specifically covered here, if recombinant import factors are used in place of extract, it is important to include BSA or other non-specific protein to reduce sticking of the import substrate to cytoplasmic structures. This is also applicable to negative controls without added extract.

[e] Typicially, 90% glycerol in PBS along with a single crystal of an antifade agent such as *p*-phenylenediamine is used as the fixative.

## 2.2 *In vivo* NLS–GFP protein import assay

This *in vivo* assay was developed to quantify the relative rates with which an NLS-green fluorescent protein (NLS-GFP) fusion protein is imported into the nucleus, in wild-type and mutant yeast cells (3, 4). The NLS-GFP fusion protein is small enough to diffuse through the NPC, however, the protein is concentrated in the nucleus in wild-type yeast due to the strong NLS. Treatment of the cells with the metabolic poison sodium azide causes the NLS-GFP to equilibrate

## Protocol 3

### *In vivo* NLS–GFP protein import assay in *Saccharomyces cerevisiae*

#### Equipment and reagents

- Fluorescent microscope equipped with 460–500 nm excitation filter and 510–560 nm emission filter using a Plan-apo ×100 oil immersion objective lens with numerical aperture of 1.35
- 25 × 200 mm glass culture tubes
- Rotator (New Brunswick Scientific, model TC-7)
- 15 ml polypropylene conical centrifuge tubes and 1.5 ml polypropylene microcentrifuge tubes
- *S. cerevisiae* strains to be used (must be *ura*$^-$ or *leu*$^-$) transformed with the NLS-GFP reporter plasmid pGADGFP *URA3* or pGADGFP *LEU2*[a]
- Synthetic dropout medium (-leucine or -uracil), autoclaved[b]
- 20% glucose in $ddH_2O$, filter sterilized
- 200 mM sodium azide in sterile $ddH_2O$ (prepared fresh)
- 200 mM 2-deoxy-D-glucose in $ddH_2O$

**Protocol 3** continued

## Method

1. Grow the *S. cerevisiae* strain to be tested in 25 × 200 mm glass culture tubes in a rotator to early log phase (0.1–0.3 $OD_{600}$) in 10 ml of synthetic dropout medium supplemented with 2% (final concentration) glucose.
2. Harvest cells in 15 ml polypropylene conical centrifuge tubes by centrifugation at room temperature at 1000 *g* for 5 min. Completely remove supernatants by aspiration.
3. Resuspend cell pellet in 1 ml of synthetic dropout medium supplemented with 10 mM sodium azide and 10 mM 2-deoxy-D-glucose. Transfer to 1.5 ml microcentrifuge tubes and gently shake for 45 min.
4. Pellet cells by centrifugation for 30 sec at room temperature. Completely remove supernatant by aspiration.
5. Wash cells with 1 ml ice-cold sterile dd$H_2O$ to remove sodium azide and 2-deoxy-D-glucose. Pellet cells by centrifugation and completely remove supernatants by aspiration. Cell pellets can be left on ice at this stage for at least 3 h.
6. At time zero, resuspend the cell pellet in 50 μl of pre-warmed synthetic dropout medium with 2% glucose and incubate in a heat block at desired assay temperature for the duration of the assay.[c]
7. Remove 2 μl of each sample at each time point (every 2–3 min) and observe under the microscope using the GFP filter.
8. Individual cells are scored as being nuclear or cytoplasmic based on the localization of NLS–GFP. A cell is scored nuclear when its nucleus fluoresces brightly enough to clearly distinguish the outline of the nucleus against the cytoplasmic background. A cell that is scored cytoplasmic may contain a relatively bright nucleus, but if the cytoplasmic background is bright enough to obscure the nuclear border, then the cell is scored cytoplasmic.
9. Score at least 50–150 cells for each time point. Relative import rates can be determined by fitting a linear regression line through the linear portion of the time-course.

[a] These plasmids are described in ref. 3.

[b] Medium used for this protocol is standard yeast growth medium (5).

[c] Standard assay temperature for wild-type yeast is 30 °C but elevated temperatures can be used for conditional mutants.

across the nuclear envelope. The rate of import is calculated as a function of time after the equilibrated cells are returned to physiological growth medium where the NLS–GFP in the cytoplasm is re-imported into the nucleus. Re-import is observed directly in living cells with a fluorescent microscope.

## 2.3 Single nuclear pore transport assay

The newly developed optical recording of signal-mediated protein transport (OSTR) through a single nuclear pore has opened the door to analysis of rates of import and/or export through individual pores (6). In this assay, the nuclear envelope of *Xenopus* oocytes is firmly attached to an isoporous filter. The system is designed such that different solutions can be placed on the nuclear versus the cytoplasmic side of the pore/envelope. A fluorescently labelled transport substrate is then added either to the nuclear or cytoplasmic side of the pore/envelope and transport across patches is quantified by confocal laser scanning microphotolysis (SCAMP).

## 2.4 *In vivo* transport studies using microinjection into *Xenopus* oocytes

Microinjection of protein and RNA into *Xenopus* oocytes and tissue culture cells have been very helpful approaches to the study of nucleocytoplasmic trafficking (7–9). A particular protein or RNA of interest can be injected into the cytoplasm of a cell and analysed for its import into the nucleus. Export from the nucleus can also be tested by injecting the sample into the nucleus. Protocol 4 focuses on microinjection into *Xenopus* oocytes, but with minor modifications this protocol can be adapted for cultured cells. The primary difference in the two approaches is that microinjection into cultured cells is generally monitored with microscopy to observe the change in localization of the microinjected protein. In contrast, since the nucleus can be physically dissected away in *Xenopus* oocytes, the analysis of these experiments tends to be through gel electrophoresis of the nuclear and cytoplasmic fractions. The biochemical nature of this type of analysis makes this system preferable for the study of trafficking of RNA molecules. Microinjections into *Xenopus* oocytes can also be helpful in visualizing the intranuclear localization of a protein.

## Protocol 4

## *Xenopus* oocyte microinjections to detect nuclear export of proteins or RNAs

### Equipment and reagents

- Dissecting microscope
- Transjector 5246 (Eppendorf)
- Motorized micromanipulator 5171 (Eppendorf)
- 18 °C incubator
- Stage V or VI *Xenopus* oocytes
- $^{35}$S-labelled methionine for protein or $^{32}$P-labelled nucleotide for RNA
- Proteinase K (Sigma)
- 50 × TAE buffer: 2 M Tris, 1 M glacial acetic acid, 0.05 M EDTA
- Phenol/chloroform/isoamyl alcohol (25:24:1)
- 70% ethanol

**Protocol 4** continued

- 6 × RNA gel loading buffer: 0.25% bromophenol blue, 0.25% xylene cyanol, 30% glycerol, DEPC-treated water
- Urea polyacrylamide denaturing gel (for RNA)
- SDS–polyacrylamide gel (for protein)
- 5 × RSB buffer: 250 mM Tris pH 6.8, 500 mM dithiothreitol (DTT), 10% SDS, 0.5% bromophenol blue, 50% glycerol

## A. Protein microinjection

1. *In vitro* translate the protein of interest and label with [$^{35}$S]methionine.
2. Inject labelled protein (in a 10 nl volume) into the nucleus of the *Xenopus* oocyte using the Transjector 5246 and motorized micromanipulator 5171.[a]
3. Appropriate controls, namely a $^{35}$S-labelled protein that is exported and a $^{35}$S-labelled protein that is not exported, should also be co-injected with the protein of interest.
4. Incubate the injected oocytes at 18 °C during the time-course.
5. At various time points, collect samples of oocytes and manually dissect the nuclei away from the oocytes.
6. Resuspend nuclear and cytoplasmic extracts in 1 × RSB buffer.
7. Resolve extracts by SDS–PAGE and analyse results via autoradiography.[b]

## B. RNA microinjection

1. *In vitro* transcribe the RNA of interest and label with [$^{32}$P]UTP.
2. Inject labelled RNA (in a 10 nl volume) into the nucleus of the *Xenopus* oocyte using the Transjector 5246 and motorized micromanipulator 5171.[c]
3. Appropriate controls, namely a $^{32}$P-labelled RNA that is exported (tRNA) and a $^{32}$P-labelled RNA that is not exported (U6 RNA), should also be co-injected with the RNA of interest.
4. Incubate the injected oocytes at 18 °C during the time-course.
5. At various time points, collect a sample of oocytes and manually dissect the nuclei away from the oocyte.
6. Resuspend nuclear and cytoplasmic extracts in 0.8 mg/ml proteinase K in 1 × TAE.[d]
7. Incubate at 37 °C for 1 h.
8. Extract the RNA from the solution using phenol/chloroform/isoamyl alcohol followed by ethanol precipitation. (Pellet should be dried without heat before proceeding.)
9. Resuspend RNA in distilled water and gel loading buffer.
10. Resolve RNA on denaturing urea/acrylamide gel and analyse via autoradiography.[e]

[a] Injection into the nucleus can be ensured by utilizing the red colour of the rabbit reticulocyte system. If the protein was successfully injected into the nucleus, the nucleus should have a red colour. If not using a rabbit reticulocyte system to translate the protein, the translated protein should be incubated in a coloured dextran solution prior to injection.

**Protocol 4** continued

[b] The level of protein export observed in the assay is followed by comparing the amount of protein in the nuclear fraction to the amount of protein in the cytoplasmic fraction throughout the time-course of the experiment.

[c] Injection into the nucleus can be ensured by incubating the transcribed RNA in a coloured dextran solution prior to injection. If the RNA was successfully injected into the nucleus, the nucleus should have the characteristic red colour.

[d] If analysing proteins bound to the RNA, omit steps 6–8. These steps strip the RNA of its proteins. Final analysis can be carried out by running an RNA or protein gel.

[e] Export of RNA from the nucleus to the cytoplasm is followed by comparing the amount of RNA in the nuclear fraction to the amount of RNA in the cytoplasmic fraction throughout the time-course of the experiment.

## 2.5 *In vivo* RNA localization using fluorescence *in situ* hybridization (FISH)

Cellular RNA molecules are synthesized in the nucleus and most need to move into the cytoplasm to perform their functions. RNA molecules can be localized *in vivo* using fluorescence *in situ* hybridization (FISH) assays. Using this method, many classes of RNA have been localized by conventional indirect immunofluorescence microscopy using digoxigenin labelled antisense probes specific for different types of RNA (e.g. poly(A)$^+$ RNA, tRNA, snoRNA, etc.). By combining this assay with conditional yeast mutants, it has been possible to define components that are critical for the intracellular trafficking of a variety of RNAs with particular emphasis on mRNA (10). The method described in Protocol 5 is the classical approach to localize poly(A)$^+$ mRNA (11). Recently variations of these methods have been adapted to detect single specific RNA messages (12) and tRNAs (13).

## Protocol 5

## Detection of poly(A)$^+$ mRNA using fluorescence *in situ* hybridization (FISH)

### Equipment and reagents

- Fluorescent microscope equipped with 460–500 nm excitation filter and 510–560 nm emission filter with a Plan-apo ×100 oil immersion objective lens with numerical aperture of 1.35
- Humidified chamber/box
- Green 10-well Teflon-coated microscope slides (CellPoint Scientific)
- 24 × 60 mm one ounce selected micro cover glasses (VWR Scientific)
- A 5 ml culture of *S. cerevisiae* cells grown to a cell density of $5 \times 10^6$ to $1 \times 10^7$ cells/ml
- Distilled or deionized water (autoclaved)
- 37% formaldehyde (Fisher)
- 0.1 M potassium phosphate buffer ($KPO_4$) pH 6.5
- P solution: 0.1 M potassium phosphate buffer pH 6.5, 1.2 M sorbitol

**Protocol 5** continued

- 1 M dithiothreitol (DTT)
- 10 mg/ml zymolyase in P solution (US Biological)
- 0.3% poly-lysine (Sigma, >150 kDa)
- 0.5% NP-40 (Sigma) in P solution, freshly prepared
- 0.1 M triethanolamine (TEA) pH 8.0, freshly prepared
- 0.25% acetic anhydride in 0.1 M TEA pH 8.0
- 20 × SSC solution: 3 M NaCl, 0.3 M sodium citrate·$2H_2O$ pH 7.0
- 50 × Denhardt's solution (Sigma)
- Pre-hybridization solution: 50% deionized formamide (Sigma), 4 × SSC, 1 × Denhardt's solution, 125 mg/ml tRNA (Sigma), 10% dextran sulfate (Sigma), 500 mg/ml salmon sperm DNA[a]
- Hybridization solution: pre-hybridization solution with digoxigenin labelled $dT_{50}$ probe
- Digoxigenin labelled probe[b]
- Antibody blocking buffer 1: 0.1 M Tris pH 9.0, 0.15 M NaCl, 5% heat-inactivated fetal calf serum (Sigma), 0.3% Triton X-100
- Anti-digoxigenin Fab–FITC (Boehringer Mannheim)
- Antibody wash 1: 0.1 M Tris pH 9.0, 0.15 M NaCl
- Antibody wash 2: 0.1 M Tris pH 9.5, 0.1 M NaCl, 50 mM $MgCl_2$
- 4,6-diamidino-2-phenylindole (DAPI) solution: 1 mg/ml in antibody wash 2
- 10 × PBS buffer: 1.5 M NaCl, 30 mM KCl, 15 mM $KH_2PO_4$, 80 mM $Na_2HPO_4 \cdot 7H_2O$ pH 7.3
- Antifade solution: phenylenediamine (a single crystal) dissolved in 1 × PBS, 90% glycerol
- Clear nail polish

## Method

1. Fix a 5 ml culture of log phase yeast cells by adding 700 μl of 37% formaldehyde. Incubate for 90–120 min on rotator at room temperature.
2. Transfer the cells to a 15 ml plastic centrifuge tube and centrifuge at 3000 rpm for 3 min at room temperature.
3. Discard the supernatant and wash the pellet twice with 0.1 M phosphate buffer and once with P solution.
4. After the final wash, resuspend the pellet in 1 ml of P solution.
5. Add DTT to a final concentration of 25 mM and incubate at room temperature for 10 min.
6. Add 50 μl of zymolyase to lyse the cell wall and incubate on rotator at room temperature for 10 min.[c]
7. During cell wall lysis, coat microscope slide wells with 0.3% poly-lysine by placing the poly-lysine solution on the slide and incubating for 5 min followed by two washes with distilled water. Air dry the slide after washing.
8. After cell lysis, collect cells at 13 000 rpm in a microcentrifuge at room temperature, discard supernatant, and resuspend cells in 1 ml of P solution.
9. Apply 15–20 μl cell suspension to each well and incubate for 15 min at room temperature.

**Protocol 5** continued

**10** Gently aspirate off excess cells by touching only the outside of the well.

**11** Permeabilize cells by treating with 0.5% NP-40 in P solution for 5 min.

**12** Aspirate off NP-40 and wash once with P solution.

**13** Equilibrate cells in freshly prepared 0.1 M TEA for 2 min at room temperature.

**14** Block polar groups by treating with 0.25% acetic anhydride/0.1 M TEA for 10 min at room temperature.

**15** Add 20 μl of pre-hybridization solution to each well and incubate slide at 37 °C for 1 h in a humidified chamber.[d]

**16** Aspirate pre-hybridization solution. Add 20 μl of hybridization solution containing probe to each well and hybridize overnight at 37 °C in a humidified chamber.

**17** Aspirate hybridization buffer and wash wells as follows: quickly in 2 × SSC; 1 h at room temperature in 2 × SSC; 1 h at room temperature in 1 × SSC; 30 min at 37 °C in 0.5 × SSC; 30 min at room temperature in 0.5 × SSC.

**18** Block cells in antibody blocking buffer for 1 h at room temperature.

**19** Add anti-digoxigenin Fab–FITC at a 1:200 dilution in antibody blocking buffer and incubate for 2 h to overnight at room temperature.[e]

**20** Aspirate secondary antibody and wash as follows at room temperature. Rinse once in antibody wash 1, and follow with 10 min and 30 min washes in the same buffer. Next, wash for 10 min in antibody wash 2, and follow with 30 min in the same buffer.

**21** Stain nuclei by treating with DAPI (1:1000 dilution) for 5 min at room temperature.

**22** Wash cells twice with antibody wash 2 for 5 min.

**23** Air dry slide.

**24** Add antifade solution to each well.

**25** Place coverslip on top of slide and seal edges with clear nail polish.

**26** View slide using fluorescence microscopy.

[a] Salmon sperm DNA should be denatured by incubating at 100 °C for 5 min and cooling on ice for at least 5 min. Denatured salmon sperm should be added to an aliquot of buffer just prior to use.

[b] The probe is critical. The oligonucleotide should be of uniform length (order a synthesized oligo(dT) 50-mer oligonucleotide, 5′ and 3′ hydroxyl, RP-1 purified). The probe is digoxigenin labelled using the Boehringer Mannheim Genius-6™ kit. Each batch of probe should be titrated for optimal signal/noise ratio.

[c] Proper cell wall digestion can be ensured by viewing cells under a phase contrast light microscope. Cells lacking cell wall will appear dark under phase microscope.

[d] Do not allow wells to dry out during hybridization

[e] At this point in the protocol a second label with another antibody can be incorporated. After incubation in the second primary antibody, the sample wells should be washed three times in antibody blocking buffer for 3 min each. The secondary antibody should then be added and incubated for 1 h at room temperature

## 3 Nucleocytoplasmic protein shuttling assays

Numerous proteins with a steady-state nuclear localization have been shown to rapidly shuttle between the cytoplasm and the nucleus (14, 15). Determining whether a nuclear protein shuttles involves determining whether the protein can be exported from the nucleus. Several assays have been developed in *S. cerevisiae* and higher eukaryotic systems to independently determine whether a nuclear protein can be exported from and re-imported into the nucleus.

### 3.1 Heterokaryon assays

Assays involving heterokaryon formation have been developed in yeast (16) and higher eukaryotes (17). Formation of heterokaryons involves cell fusion in the absence of nuclear fusion to form a cytoductant. This can be accomplished by fusing tissue culture cells from different species (Figure 1A) or in *S. cerevisiae* by mating a wild-type cell with a *kar1-1* mutant (16, 18) such that nuclear fusion does not occur (Figure 1B). In either case, one of the cell types to be fused expresses the nuclear protein of interest and the other does not. The goal of the assay is to determine whether that nuclear protein can exit the first nucleus and enter the second nucleus in the absence of ongoing new protein synthesis.

**Protocol 6**

### Analysis of protein shuttling in mammalian heterokaryons

#### Equipment and reagents

- Fluorescent microscope with Plan-apo oil immersion ×60 objective equipped with appropriate filters to detect your protein of interest
- Tissue culture equipment
- 18 mm$^2$ glass coverslips
- Cultured NIH 3T3 cells transfected with a tagged version of the protein to be analysed
- Immunofluorescence materials
- Cultured HeLa cells (untransfected)
- Trypsin
- Tissue culture medium
- 100 μg/ml cycloheximide
- Phosphate-buffered saline (PBS)
- Polyethylene glycol 5000

#### Method

1. Grow NIH 3T3 cells to subconfluent levels.
2. Harvest the cells using trypsinization and seed the cells onto 18 mm$^2$ glass coverslips.
3. Once the cells are attached seed an equal number of HeLa cells onto the coverslips.
4. Incubate the co-cultures until the second set of cells have attached to the coverslip (approx. 3–4 h).
5. Incubate the co-cultures with 100 μg/ml cycloheximide for 30 min to inhibit new protein synthesis.
6. Wash the coverslips with PBS.[a]
7. Fuse cells by inverting the coverslips onto a drop of pre-warmed (at 37 °C) polyethylene glycol 5000/PBS for 120 sec.[b]
8. Wash the coverslips with PBS.

**Protocol 6** continued

**9** Incubate the fused cells with 100 μg/ml cycloheximide for 60 min.

**10** Fix the cells for indirect immunofluorescence and process by standard methods to detect your protein of interest.[c]

[a] Alternatively, washes of the cultures can also be performed with buffers that are specific for the specific cell types.

[b] After the fusion step, addition of 10 mM cytosinarabinoside to the fused co-cultures prevents overgrowth of non-fused cells.

[c] In many cases it may be possible to use a GFP tagged protein so that cells can be viewed directly through fluorescence of the GFP molecule; however, cells are generally fixed for at least some experiments so they can be co-stained with a DNA dye such as Hoescht to visualize nuclei.

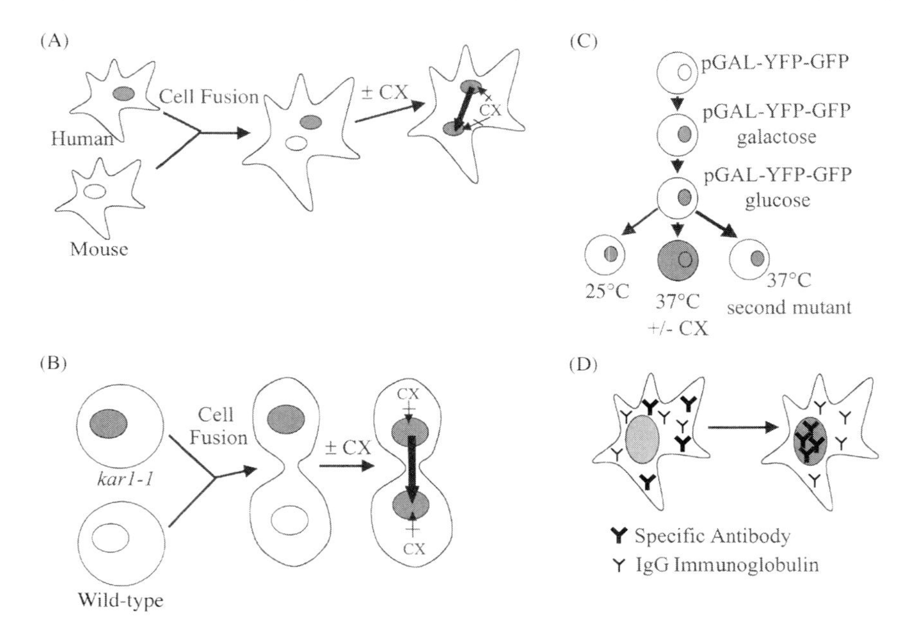


**Figure 1** Schematic representation of various mammalian and yeast shuttling assays. (A) Heterokaryon formation and shuttling using human and mouse cell lines. Nuclear proteins that shuttle are detected in both nuclei of the heterokaryon whereas non-shuttling nuclear proteins are detected in a single nucleus. (B) Heterokaryon formation and shuttling in *S. cerevisiae* using a wild-type strain and a *kar1–1* strain that is defective in nuclear fusion. Nuclear proteins that shuttle are detected in both nuclei whereas non-shuttling nuclear proteins are detected in a single nucleus. (C) Export of nuclear proteins in a *nup49–313* mutant *S. cerevisiae* strain. Following transient expression of the GFP tagged protein of interest and shift to the non-permissive temperature, nuclear proteins that shuttle appear throughout the cell whereas non-shuttling nuclear proteins are restricted to the nucleus. (D) Identification of shuttling proteins by nuclear import of specific antibodies. Following injection of specific antibodies into the cytoplasm, nuclear proteins that shuttle will bind to their specific antibodies and import them into the nucleus whereas antibodies against non-shuttling nuclear proteins will remain in the cytoplasm. CX = cycloheximide.

## 3.2 *nup49* nuclear protein export assay in *S. cerevisiae*

In *S. cerevisiae* it is possible to determine whether a protein can exit the nucleus by using a conditional mutant cell that is able to export nuclear proteins but is unable to carry out protein import following a shift to non-permissive conditions. In these cells the protein of interest accumulates in the cytoplasm if the protein is able to exit the nucleus (Figure 1C). This has been accomplished by taking advantage of a specific mutation in a nuclear pore protein, *nup49–313*, that blocks protein import but not protein export (19). Briefly, the protein of interest (your favourite protein—YFP) is tagged with green fluorescent protein (YFP–GFP) and is expressed under the control of an inducible promoter for a few hours and then rapidly repressed such that the only proteins present in the cell have already been targeted to the nucleus. This assay will only be informative if YFP is completely nuclear at steady-state and its import is blocked in the *nup49–313* mutant. The cells are then shifted to the non-permissive temperature and the localization of YFP–GFP is observed. The detailed methodology is presented in Protocol 7. One advantage of this method is that the effect of protein localization in double mutants containing *nup49–313* and other mutants can be examined to determine whether the second gene is required for the export of your protein of interest (20).

## Protocol 7

# Export of shuttling proteins from the nucleus in *S. cerevisiae* using the *nup49–313* nucleoporin allele

### Equipment and reagents

- Fluorescent microscope equipped with excitation filter of 460–500 nm and an emission filter 510–560 nm using a Plan-apo ×100 oil immersion objective lens with numerical aperture of 1.35
- 25 × 75 × 1 mm frosted microscope slides
- 24 × 40 mm micro coverslips
- *nup49–313* yeast strain transformed with a plasmid encoding your protein of interest fused to GFP and expressed from the inducible *GAL1–10* promoter[a,b]
- 20% glucose
- 20% galactose
- Dropout yeast growth medium[c]
- 20% raffinose
- Deionized water

### Method

1 Grow the *nup49–313* cells transformed with your shuttling proteins in dropout medium containing 2% glucose to saturation at 25 °C.[d]

**Protocol 7** continued

2 Dilute (1:200 dilution) and grow the cells in dropout medium containing 2% raffinose at 25 °C.

3 When the cells reach $1 \times 10^7$ cells/ml, add 2% galactose to the cultures and incubate at 25 °C for 3 h.

4 Wash the cells once with $dH_2O$ and resuspend the cells in dropout medium containing 2% glucose.

5 Incubate the cells at 25 °C for 2 h.

6 Split each culture and incubate one-half of the culture at 25 °C and the other half of the culture at 37 °C for 5 h.

7 Remove 1 ml of cells from each culture, wash once with $dH_2O$, and resuspend in 25–100 μl of $dH_2O$.

8 Place 3 μl of washed cells onto a glass slide and analyse nuclear export by fluorescence microscopy.[e]

[a] Cells should also be transformed with plasmids encoding control shuttling (Np13p–GFP) and non-shuttling proteins (NLS–LacZ–GFP) (19).

[b] Alternatively, cells expressing your protein–GFP fusion under the control of its endogenous promoter (pYFP) can be used in the assay. Steps 2–6 should be modified to include growing the cells for 2–3 h until they are in mid log phase followed by incubation of the cells with 100 μg/ml cycloheximide before splitting and shifting the cells.

[c] All yeast growth medium is prepared by standard protocols (5).

[d] This assay can be coupled with other mutants to examine whether other proteins affect nuclear export of YFP–GFP.

[e] The protein of interest has been exported from the nucleus to the cytoplasm when diffuse cytoplasmic staining is observed. Often there is still significant nuclear staining as all of the protein has not been exported from the nucleus, but it is the mere appearance of signal in the cytoplasm that can be used, in conjunction with the proper control proteins, to confirm nucleocytoplasmic shuttling.

### 3.3 Antibody injection to detect protein shuttling

Finally, shuttling in higher eukaryotic cells can be performed in single cells using the import of specific antibodies to nuclear antigens (Figure 1D). Specific and non-specific immunoglobulins are injected into the cytoplasm of mammalian cells and their localization patterns are detected by immunofluorescence. This procedure is dependent on the availability of the nuclear antigen in the cytoplasm after export from the nucleus. The specific antibody binds to the nuclear antigen in the cytoplasm and is imported into the nucleus as an antibody/antigen complex. The disadvantage of this procedure is the probability that the nuclear localization signal of the nuclear antigen may be hidden when complexed with the antibody and thus re-import may not occur.

# 4 Regulated nuclear transport: protein–protein interactions

Three basic strategies for monitoring protein–protein interactions are described here. The following general approaches are described:

(a) Immunological techniques using antibodies (Protocol 8).

(b) Techniques using radiolabelling nucleotides (Protocols 9 and 10).

(c) Use of fluorescent proteins to detect molecular interactions *in vivo* and *in vitro*.

Immunological and radiolabelling techniques have been very convenient approaches to identify receptors and their ligand binding characteristics. The methods most used in the field of nucleocytoplasmic transport involve the application of antibodies against the ligand or the use of radiolabelled ligand to identify the binding partner after separation on polyacrylamide gels and blotting on a nitrocellulose membrane (see for example refs 21–23). The advantage of these methods is based on renaturation and recovery of ligand binding activity to better establish the properties of the ligand binding to the protein complex.

## 4.1 Ligand binding studies using overlay assays

### 4.1.1 Overlay assays using immunodetection

One of the most used methods to monitor the functional interactions between proteins involved in nuclear transport is a blot overlay assay using 'immunostaining' or 'immunoblotting'. This type of assay serves as a sensitive monitor for *in vitro* protein–protein interactions. This approach has been particularly useful for studying interactions between soluble nuclear transport factors and components of the nuclear pore (21–23). Nuclear pore proteins (nucleoporins) are extremely hydrophobic and have been notoriously difficult to express and purify. Thus, blot overlay assays are particularly useful as binding to components of the nuclear pore can be studied without the need for purified preparations of nucleoporins. Recently this type of assay was used to study the interaction between a nucleoporin and the transport receptor importin-beta (21). The assay was performed with an S-tagged fusion protein using the S-Tag™ purification kit from Novagen (Novagen). The use of this system enables analytical to medium scale affinity purification of the target protein and its subsequent immunodetection using S-protein HRP (horseradish peroxidase) antibody (24).

## Protocol 8

## Blot overlay assay to detect protein–protein interactions

### Equipment and reagents

- Electrophoresis and electrotransfer apparatus
- Filter paper (Whatman No. 1)
- Nitrocellulose membrane (Schleicher & Schuell)
- Electrophoresis running buffer: 0.192 M glycine, 0.025 M Tris, 0.1% (w/v) SDS
- Transfer buffer: 25 mM Tris–HCl pH 8.3, 192 mM glycine, 20% (v/v) methanol
- Blocking buffer: 4% Marvel (fat-free milk powder) in PBS containing 2 mM dithiothreitol (DTT)
- Phosphate-buffered saline (PBS)
- Protein sample buffer: 50 mM Tris–HCl pH 6.8, 10% (v/v) glycerol, 2% (w/v) glycerol, 2% (w/v) SDS, 100 mM DTT, 0.1% bromophenol blue
- Purified recombinant S-tagged protein
- Cell lysate (or purified protein) samples (frozen) for analysis
- S-protein HRP (Novagen)
- Chemiluminescence substrate: ECL developing reagents (Pharmacia)

### Method

1. Add protein sample buffer to cell lysate (or purified protein) sample and boil for 5 min. Vortex and centrifuge the tubes in a microcentrifuge for 1 min to pellet any insoluble material. Carefully remove the supernatant for analysis.
2. Load 5–10 μg total protein sample onto gel and electrophorese the samples towards the positive electrode at 150 V for 1–2 h, until the bromophenol blue dye front approaches the bottom of the gel.
3. Separate the gel plates, remove the gel using gloved hands, and build up the transfer stack in the cassette for blotting one layer at a time. Keep each layer fully submerged below the transfer buffer. In order, the layers are: one sponge pad, a sheet of filter paper, the nitrocellulose membrane, the SDS–PAGE gel, a sheet of filter paper, the second sponge pad. Exclude air bubbles from between the layers by using a glass rod or test-tube as a roller to smooth each layer and expel visible air bubbles.
4. Close the assembly cassette and insert it in the blotting tank oriented with the nitrocellulose membrane on the anodal side. Pour cold (4 °C) transfer buffer in the tank, insert a magnetic stirrer bar and a cooling block. Place the tank on a magnetic stirrer and run at 100 V for 1 h.
5. Check the efficiency of transfer by visualization of the pre-stained molecular weight markers.
6. Incubate the blot in 20 ml of blocking buffer for 30 min at room temperature and 30 min at 4 °C with constant shaking.
7. Wash the blot for 5 min in PBS buffer.
8. Overlay the membrane with 4 ng/ml purified S-tagged protein in PBS buffer containing 0.2% Tween 20 for 1–2 h at room temperature.
9. Wash the membrane as described in step 7.

**Protocol 8** continued

**10** Dilute the S-protein HRP antibody to 1:5000 in PBS containing 0.2% Tween 20. Pour this onto the membrane and incubate at room temperature for 1 h.

**11** Wash the membrane as described in step 7.

**12** When the blot has been washed, pour the ECL developing solution and leave it for 1 min at room temperature with constant shaking.

**13** Discard the developing solution, dry the blot between two layers of filter paper, and immediately take the blot to the dark-room.

**14** Working with only a dark-room safe light for illumination, place a sheet of X-ray film on top of the blot. Leave 5 sec to 10 min depending on the intensity of the chemiluminescence.

**15** Develop the X-ray film following manufacturer's instructions.

### 4.1.2 Radiolabelling method: detection of GTP binding proteins and proteins that interact with GTP binding proteins

A modification of the overlay assay using photoaffinity derivatives of GTP such as [$\gamma$-$^{32}$P]GTP azidoanidile has been very useful for *in vitro* assays to identify the constituents of the Ran GTPase system. The Ran system regulates the loading of the cargo onto nuclear transport receptors (25). In general, this technique can be of use in identifying a variety of GTP binding proteins (Protocol 9). It has been shown that different types of photoaffinity derivatives of GTP can be used to label specific classes of GTP binding proteins. [$\alpha$-$^{32}$P]GTP itself, for example, may be useful to label only GTP binding proteins in the 25 kDa molecular weight range, after these have been electroblotted (26). Under these conditions the 40 kDa molecular weight proteins are not labelled and only the 20–30 kDa proteins are visualized. In Protocol 10, the same method, with some minor modifications, is used to monitor the binding of Ran to other nuclear transport proteins in a nucleotide-dependent manner (27, 28).

## Protocol 9

## Modified overlay blot to detect GTP binding proteins

### Equipment and reagents

- Electrophoresis and electrotransfer apparatus
- Geiger counter
- Electrophoresis running buffer: 0.192 M glycine, 0.025 M Tris, 0.1% (w/v) SDS
- Transfer buffer: 25 mM Tris–HCl pH 8.3, 192 mM glycine, 20% (v/v) methanol
- GTP binding buffer: 50 mM Tris–HCl pH 7.5, 50 mM $MgCl_2$, 2 mM dithiothreitol (DTT), 0.3% Tween 20, 0.3% bovine serum albumin, 0.1 mM ATP[a]
- GTP binding buffer plus [$\gamma$-$^{32}$P]GTP: add [$\gamma$-$^{32}$P]GTP to the GTP binding buffer to a specific activity of 5000 cpm/fmol GTP

**Protocol 9** continued

### Method

1 Prepare and run the protein sample on a polyacrylamide gel as described in Protocol 8.
2 Separate the gel plates, remove the gel using gloved hands, and electrotransfer the protein sample at 100 V for 1 h (Protocol 8).
3 Check the efficiency of transfer by visualization of the pre-stained molecular weight marker.
4 Renature the protein sample by incubating the membrane in GTP binding buffer at room temperature for 1 h.[b]
5 Incubate the membrane with GTP binding buffer plus [$\gamma$-$^{32}$P]GTP for 1 h at room temperature.
6 Wash the membrane with GTP binding buffer for 5 min at least three times. Monitor loss of background radioactivity with a Geiger counter.
7 Autoradiograph membrane by placing a sheet of X-ray film on top of the blot and by leaving it at $-70\,^{\circ}$C overnight.
8 Develop the X-ray film following manufacturer's instructions.

[a] This solution should be filtered through an Amicon 0.45 μm syringe tip filter unit.

[b] The incubation temperature and time may be optimized for your protein sample.

## Protocol 10

## Modified overlay blot to identify Ran binding proteins

### Equipment and reagents

- See Protocol 9[a]
- Purified recombinant Ran protein

### Method

1 Electrophorese and electroblot the protein sample as described in Protocol 8.
2 Check the efficiency of transfer by visualization of the pre-stained molecular weight marker.
3 Renature the protein sample by incubating the membrane in GTP binding buffer at room temperature for 1 h.[b]
4 Incubate the membrane with GTP binding buffer containing 100 mM of unlabelled GTP for 30 min at 25 °C. Add 100 μl of 1 nM Ran [$\gamma$-$^{32}$P]GTP and incubate for another 10 min.
5 Wash the membrane with GTP binding buffer for 5 min at least three times. Monitor loss of background radioactivity with a Geiger counter.

**Protocol 10** continued

6 Autoradiograph membrane by placing a sheet of X-ray film on top of the blot and by leaving it at −70 °C overnight.

7 Develop the X-ray film following manufacturer's instructions.

[a] The GTP binding buffer should be filtered using an Amicon 0.45 μm syringe tip filter unit.

[b] The incubation temperature and time may be optimized for your protein sample.

## 4.2 Fluorescence-based applications to map protein–protein interactions

In the few years since the gene was first cloned (29), the *Aequorea victoria* green fluorescent protein (GFP) has become a powerful tool for studying nucleo-cytoplasmic transport (30). Several fluorescently tagged transporters as well as cargoes have been used as markers for protein localization and protein dynamics in living cells. Recently a GFP-based variation of a steady-state fluorescence depolarization was developed to examine quantitatively the regulatory mechanism of nucleocytoplasmic transport. This *in vitro* assay provides a useful method for monitoring protein–protein interactions (31). Another very successful method for detecting molecular interactions within living cells involves fluorescence resonance energy transfer (FRET) between spectral variants of GFP. This technique has been applied as a powerful intracellular molecular sensor for the *in vivo* study of nuclear transport and dissection of the nuclear pore complex (32).

### 4.2.1 Steady-state fluorescence depolarization

Fluorescence polarization is used extensively as a technique to measure protein–protein interactions (33). This method generally takes advantage of the intrinsic fluorescence of proteins and relies on a change in the depolarization of fluorescent incident light upon ligand binding. Unfortunately, this approach can be limited by the molecules being studied as one or the other of the molecules must have a sufficient fluorescent signature such that a change in depolarization can be observed following binding. In the past this has presented a problem for using this approach to study some of the protein-protein interactions that are critical to nuclear transport. This has been particularly true for the nuclear localization sequence (NLS) receptor, which contains too many tryptophans to be amenable to this approach. Thus, a modification of this procedure has been developed and applied to study the interaction between the NLS receptor and an NLS-containing substrate (31). This method takes advantage of the intrinsic fluorescence of GFP as the NLS is fused to the GFP protein and binding of this substrate to the receptor is readily observed in a standard fluorescence depolarization assay. Recently this method was used to measure the binding affinity between an NLS tagged with GFP and its import receptor (31). This type of assay helped not only in the determination of the relative affinity of conserved sequences for their receptor, but also the analysis of the interactions that occur during a nuclear transport event. One important aspect of this technique is the possibility in correlating an interaction energy measured *in vitro*, like the binding of nuclear localization sequences to receptor, with the functional consequences of this interaction *in vivo*. In this regard the assay can provide a quantitative analysis to develop a detailed thermodynamic model for the mechanism of a biological pathway.

### 4.2.2 Using GFP in fluorescence resonance energy transfer (FRET)-based applications

Considerable progress has been made in determining the *in situ* spatio-temporal dynamics of the molecular interactions required for nuclear transport in living cells. Lately, the use of GFP in fluorescence resonance energy transfer (FRET) has generated powerful tools that serve as genetically encoded sensors within living cells. This method is based on the fluorescence resonance energy transfer between spectral variants of GFP (34). When two fluorophores are in sufficient proximity (<100 Å) and an appropriate relative orientation such that an excited fluorophore (donor) can transfer its energy to a second, longer-wavelength fluorophore (acceptor) in a non-radiative manner. Thus, excitation of the donor can produce light emission from the acceptor, with attendant loss of emission from the donor. Because FRET is a non-destructive spectroscopic method for measuring molecular interactions, it can be carried out in living cells. Recently, this technique has been successfully used to investigate transport receptor–nucleoporin interactions in living cells (32). This will continue to be a valuable approach to measure protein–protein interactions *in vivo* and may be particularly useful for characterizing interactions with nuclear pore proteins due to the inherent difficulties in studying this class of proteins using classical biochemical approaches.

## 5 Genetic approaches to studying nuclear transport

Numerous advances in our understanding of nucleocytoplasmic transport processes have arisen from the characterization of conditional mutants defective in some aspect of transport and the use of these alleles as the basis for genetic screens. This has been particularly true for the budding yeast, *S. cerevisiae*. For example, components of the Ran pathway were originally identified in this organism in the 1960s (35). Ran itself was identified in yeast as a suppressor of a mutation in the Ran exchange factor (36). More recently yeast genetics has been used extensively and very successfully to identify numerous components of the nuclear pore through a large number of genetic screens (see ref. 37 for review). Thus, an excellent approach to identifying novel components of the nucleocytoplasmic transport machinery or to studying interactions between known components is to use standard yeast genetic screens and assays (38).

## References

1. Adam, S. A., Sterne-Marr, R., and Gerace, L. (1992). In *Methods in enzymology*, Vol. 219, pp. 97–110.
2. Smythe, C. and Newport, J. W. (1991). *Methods Cell Biol.*, **35**, 449.
3. Shulga, N., Roberts, P., Gu, Z., Spitz, L., Tabb, M. M., Nomura, M., *et al.* (1996). *J. Cell Biol.*, **135**, 329.
4. Roberts, P. M. and Goldfarb, D. S. (1998). *Methods Cell Biol.*, **53**, 545.
5. Adams, A., Gottschling, D. E., Kaiser, C. A., and Stearns, T. (1997). *Methods in yeast genetics*. Cold Spring Harbor Laboratory Press, Cold Spring Harbor.
6. Keminer, O., Siebrasee, J. P., Zerf, K., and Peters, R. (1999). *Proc. Natl. Acad. Sci. USA*, **96**, 11842.
7. Nemergut, M. E. and Macara, I. G. (2000). *J. Cell Biol.*, **149** (4), 835.
8. Bellini, M. and Gall, J. G. (1999). *Mol. Biol. Cell*, **10** (10), 3425.
9. Terns, M. P. and Goldfarb, D. S. (1998). *Methods Cell Biol.*, **53**, 559.
10. Cole, C. N. (2000). *Nature Cell Biol.*, **2**, E55.
11. Amberg, D. C., Goldstein, A. L., and Cole, C. N. (1992). *Genes Dev.*, **6**, 1173.
12. Saavedra, C. A., Hammell, C. M., Heath, C. V., and Cole, C. N. (1997). *Genes Dev.*, **11**, 2845.
13. Sarkar, S. and Hopper, A. K. (1998). *Mol. Biol. Cell*, **9**, 3041.
14. Piñol-Roma, S. and Dreyfuss, G. (1992). *Nature*, **335**, 730.
15. Michael, W. M. (2000). *Trends Cell Biol.*, **10** (2), 46.
16. Flach, J., Bossie, M., Vogel, J., Corbett, A. H., Jinks, T., Willins, D. A., *et al.* (1994). *Mol. Cell. Biol.*, **14** (12), 8399.
17. Borer, R. A., Lehner, C. F., Eppenberger, H. M., and Nigg, E. A. (1989). *Cell*, **56** (3), 379.
18. Rose, M. D. and Fink, G. R. (1987). *Cell*, **48** (6), 1047.
19. Lee, M. S., Henry, M., and Silver, P. A. (1996). *Genes Dev.*, **10**, 1233.
20. Shen, E. C., Henry, M. F., Weiss, V. H., Valentini, S. R., Silver, P. A., and Lee, M. S. (1998). *Genes Dev.*, **12**, 679.
21. Bayliss, R., Littlewood, T., and Stewart, M. (2000). *Cell*, **102**, 99.
22. Lane, C. M., Cushman, I., and Moore, M. S. (2000). *J. Cell Biol.*, **151**, 321.
23. Rexach, M. and Blobel, G. (1995). *Cell*, **83**, 683.
24. Ho, I. C., Hodge, M. R., Rooney, J. W., and Glimcher, L. H. (1996). *Cell*, **85** (7), 973.
25. Görlich, D. (1999). *Annu. Rev. Cell Biol.*, **15**, 607.
26. Bhullar, R. P. and Haslam, R. J. (1987). *Biochem. J.*, **245** (2), 617.
27. Kutay, U., Izzaurralde, E., Bischoff, F. R., Mattaj, I. W., and Görlich, D. (1997). *EMBO J.*, **16**, 1153.
28. Bischoff, F. R., Krebber, H., Smirnova, E., Dong, W., and Ponstingl, H. (1995). *EMBO J.*, **14**, 705.
29. Chalfie, M., Tu, Y., Euskirchen, W., Ward, W., and Prasher, D. C. (1994). *Science*, **263**, 802.
30. Kahana, J. A. and Silver, P. A. (1996). In *Current protocols in molecular biology* (ed. F. M. Ausubel, R. Brent, R. E. Kingston, D. E. Moore, J. G. Seidman, J. A. Smith, and K. Struhl), Vol. 1, pp. 9.6.13–9.6.19. John Wiley and Sons, Inc., New York.
31. Fanara, P., Hodel, M. R., Corbett, A. H., and Hodel, A. E. (2000). *J. Biol. Chem.*, **275**, 21218.
32. Damelin, M. and Silver, P. A. (2000). *Mol. Cell*, **5**, 133.
33. LeTilly, V. and Royer, C. A. (1993). *Biochemistry*, **32** (30), 7753.
34. Heim, R. and Tsien, R. Y. (1996). *Curr. Biol.*, **6** (2), 178.
35. Hartwell, L. (1967). *J. Bacteriol.*, **93**, 1662.

36. Belhumeur, P., Lee, A., Tam, R., DiPaolo, T., Fortin, N., and Clark, M. W. (1993). *Mol. Cell. Biol.*, **13**, 2152.
37. Doye, V. and Hurt, E. (1997). *Curr. Opin. Cell Biol.*, **9**, 401.
38. Appling, D. R. (1999). *Methods*, **19** (2), 338.

# Chapter 7
# Transport across the membrane of the endoplasmic reticulum

Karin Römisch
Cambridge Institute for Medical Research and Department of Clinical Biochemistry, Wellcome Trust/MRC Building, Addenbrooke's Hospital, Hills Road, Cambridge CB2 2XY, UK.

## 1 Introduction

This chapter describes cell-free experimental systems designed to monitor both protein translocation across the endoplasmic reticulum (ER) membrane and the subsequent fates of the translocated proteins. The cell fractions for the assays described here are derived from the yeast *Saccharomyces cerevisiae*. These assays can therefore be performed with membranes or cytosol from mutant yeast strains, and thus allow the assessment of the roles of specific genes in translocation across the ER membrane in addition to the biochemical manipulations that are possible in all cell-free systems. Similar experimental systems based on mammalian microsomes and cytosol have been described elsewhere (1, 2).

## 2 Translocation into the ER and monitoring import

Protein translocation across the ER membrane was the first protein transport event which was reproduced in a cell-free assay (3). These original assays were based on purified poly(A) mRNA which was translated in rabbit reticulocyte or wheat germ lysate and translated proteins were translocated into membrane vesicles derived from tissues rich in ER and low in intrinsic RNase activity, such as chicken oviduct or dog pancreas. While the mRNA used in modern ER translocation assays is generally transcribed *in vitro* (4), reticulocyte/wheat germ lysate and dog pancreas microsomes are still widely used and now commercially available (Promega). Laboratories working with yeast generally prepare their own translation extracts and microsomes from wild-type and mutant yeast strains (4, 5). Most use protocols that are similar to those described here. The preparation of yeast S-100 translation extract is described in Protocol 1 and the use of this extract is described in Protocol 2.

## Protocol 1

# Preparation of yeast S-100 translation extract

### Equipment and reagents

- Autoclaved glass column with 300 ml bed volume; autoclaved connecting tubing
- Autoclaved stainless steel buckets (1 litre) for blender, electric blender
- Autoclaved 50 ml centrifuge tubes
- Autoclaved 45Ti tubes
- Medium speed centrifuge; SS34 rotor or equivalent
- Ultracentrifuge; 45Ti rotor or equivalent
- Liquid nitrogen
- Autoclaved 300 ml hydrated G-25 Sephadex (medium or coarse) in DEPC-water
- 10 litres YPD: 10 g/litre bacto-yeast extract (Difco), 20 g/litre bacto-peptone (Difco); after autoclaving, add glucose to 2% (w/v) from a sterile 50% (w/v) stock
- 5 litres diethyl pyrocarbonate(DEPC)-treated, deionized, sterile water: add 1 ml DEPC (Sigma) per 1 litre of water, mix; incubate the water at room temperature for 4 h to overnight, then autoclave it for 20 min
- 2 litres of 2 × buffer A: 200 mM potassium acetate, 4 mM magnesium acetate, 40 mM Hepes–KOH pH 7.4; treat the buffer with DEPC as described above
- 1 litre DEPC-treated 70% (v/v) glycerol
- 1 × buffer A: mix equal volumes DEPC-treated 2 × buffer A, and DEPC-treated water; add RNase-free DTT to 2 mM
- 1 × buffer A/14% glycerol: mix 500 ml 2 × buffer A, 200 ml of 70% DEPC-treated glycerol, and 300 ml DEPC-treated water; add RNase-free DTT to 2 mM
- 100 mM PMSF in ethanol

### Method

1. Grow up10 litres of a vacuolar protease-deficient strain (for example GPY60) in YPD to 3–5 $OD_{600}$/ml
2. Harvest the cells by centrifugation at 5000 g for 5 min, and wash the cells with 2 litres of deionized water.
3. Collect the cell pellet (approx. 40 g of cells) and wash the cells with 500 ml of DEPC-treated water.
4. Resuspend the pellet in a small volume (10 ml or less) of cold, RNase-free buffer A containing 2 mM DTT.
5. Place a plastic beaker containing liquid nitrogen and a large stir bar onto a stir plate; pour the cells into the stirring liquid nitrogen in a thin stream; this results in small frozen cell pellets. Avoid forming large clumps. The cells can be stored frozen at −80 °C for several months.
6. In the cold room, pour liquid nitrogen into the blender, add the frozen cells, and blend continuously at high speed for 10 min to lyse the cells. During this period, keep adding nitrogen through the central hole in the lid to maintain the level just below the rim of the blender.
7. Pour the slurry into a sterile beaker and allow the nitrogen to evaporate. Thaw the lysed cells in a 20 °C water-bath. Add 40 ml of 1 × buffer A containing 2 mM DTT, 1 mM PMSF.

**Protocol 1** continued

8. Transfer the cell lysate to 50 ml centrifuge tubes. Pellet unbroken cells and large fragments by centrifugation at 10 000 g for 10 min at 4 °C.
9. Transfer the supernatant to one or two 45Ti tubes; sediment membranes at 100 000 g for 35 min at 4 °C.
10. Collect the supernatant avoiding the lipid layer on top.
11. Measure $A_{260}$ of a 1:200 dilution in water.
12. Pack the Sephadex column and equilibrate it in the cold room with 2 column volumes of buffer A/14% glycerol/2 mM DTT.
13. Load the S-100 onto the column and let it run into the column.
14. Reattach the buffer supply and adjust the flow rate to 1 ml/min. Start collecting 2 ml fractions. The desired peak coincides with the first yellowish band.
15. Measure the $A_{260}$ of a 1:200 dilution in water of each fraction.
16. Pool the fractions with an $A_{260} > 30$. The $A_{260}$ of the pool should in the range of 60–100.
17. Divide the pool into convenient aliquots (0.5–1 ml) in RNase-free tubes, freeze in liquid nitrogen, and store at −80 °C. A typical preparation will yield about 30 ml.

## Protocol 2

# *In vitro* translation in yeast S-100

### Equipment and reagents

- RNase-free microcentrifuge tubes and pipette tips
- Whatman 3MM filter paper discs
- Thermoblock
- Scintillation vials, scintillation fluid, scintillation counter
- S7 micrococcal nuclease (Boehringer): 20 000 U/ml in 1 × translation buffer; frozen in 20 μl aliquots and stored at −80 °C
- 40 mM calcium chloride, DEPC-treated
- 100 mM EGTA, DEPC-treated
- RNasin (human placenta RNase inhibitor, Amersham)
- Creatine phosphokinase (Boehringer): 10 mg/ml in 50% DEPC-treated glycerol
- DEPC-treated water
- $[^{35}S]$methionine, *in vitro* translation grade (Amersham SJ1515)
- Prepro-α-factor mRNA
- 3 × translation buffer: 75 mM creatine phosphate (Boehringer), 2.25 mM ATP, 300 μM GTP, 100 μM amino acids without methionine (from 1 mM stock, Promega), 360 mM potassium acetate, 6 mM magnesium acetate, 66 mM Hepes–KOH pH 7.4, 5.1 mM DTT; all solutions must be RNase-free. The buffer should be frozen in liquid nitrogen in aliquots and only thawed once.
- Trichloroacetic acid (TCA)

**Protocol 2** continued

### Method

1 In an RNase-free tube, mix 180 μl S-100 extract, 5 μl micrococcal nuclease (20 000 U/ml), and 2.5 μl of 40 mM calcium chloride. Incubate for 20 min at 20 °C.

2 Add 2.5 μl of 100 mM EGTA. Incubate for 5 min at 20 °C.

3 In an RNase-free tube, combine 180 μl nuclease-treated S-100, 180 μl of 3 × translation buffer, 150 U RNasin, 10.8 μl creatine phosphokinase (10 mg/ml), 50 μg mRNA in DEPC-treated water, and 300 μCi [$^{35}$S]methionine. Add DEPC-treated water to 540 μl. Mix the solution by pipetting.

4 Incubate the translation reaction for 50 min at 20 °C. Transfer the translation to ice, and freeze it in aliquots in liquid nitrogen. Free [$^{35}$S]methionine and ATP can be removed by gel filtration prior to freezing.

5 To measure incorporation, spot 2 × 2 μl translation onto Whatman 3MM filter paper discs, and incubate the discs in 10% (w/v) TCA on ice for 10 min. Transfer the discs to 5% (w/v) TCA at room temperature for 5 min, then wash the discs twice in 96% (w/v) ethanol and air dry. Place the dry discs in scintillation vials, add the appropriate scintillation fluid, and count. The typical incorporation for prepro-α-factor (18 kDa, five methionines) ranges from 200 000–500 000 cpm per 2 μl translation.

The preparation of yeast microsomes is described in Protocol 3.

## Protocol 3

## Preparation of yeast microsomes

### Equipment and reagents

- Medium speed centrifuge; SS34 rotor or equivalent
- Motor-driven glass/Teflon homogenizer with tight-fitting pestle, 50 ml vessel volume
- 5 ml glass/Teflon homogenizer
- Ultracentrifuge; SW55 Ti rotor or equivalent
- YPD (see Protocol 1)
- Zymolyase 100T (Seikagaku)
- Zymolyase buffer: 0.7 M sorbitol, 0.7 × YP, 0.5% glucose, 50 mM Tris–HCl pH 7.5, 10 mM DTT
- 2 × lysis buffer: 0.4 M sorbitol, 100 mM potassium acetate, 4 mM EDTA, 40 mM Hepes–KOH pH 7.4
- 100 mM PMSF in ethanol
- 1 M DTT
- B88: 20 mM Hepes–KOH pH 6.8, 250 mM sorbitol, 150 mM potassium acetate, 5 mM magnesium acetate
- Sucrose solutions: 20 mM Hepes–KOH pH 7.4, 50 mM potassium acetate, 2 mM EDTA; 1.2 M or 1.5 M sucrose; add 1 mM DTT, 1 mM PMSF just prior to use

**Protocol 3** continued

## Method

1. Grow the cells in YPD or selective medium as required to $OD_{600}$ = 4 (if grown in YPD, microsomes for import into the ER only), or $OD_{600}$ = 1–2 (if grown in minimal medium or if microsomes are going to be used in ER degradation assays).
2. Harvest the cells for 3 min at 5000 g at room temperature.
3. Resuspend the cells to 100 $OD_{600}$/ml in freshly made 100 mM Tris-HCl pH 9.4, 10 mM DTT. Incubate the cells for 10 min at room temperature.
4. Harvest the cells as in step 2.
5. Resuspend the cells to 100 $OD_{600}$/ml in zymolyase buffer. Add 1 mg Zymolyase 100T per 1000 OD cells, and incubate at 30 °C or 24 °C (temperature-sensitive mutants) until the $OD_{600}$ of a 1:200 dilution in water is 10% of the original value, but no longer than 30 min.
6. Chill the cells on ice for 2 min.
7. Harvest the cells at 5000 g for 5 min at 4 °C.
8. Gently resuspend the spheroplast pellet in 2 × lysis buffer to about 250 $OD_{600}$/ml.
9. Harvest the cells at 5000 g for 5 min at 4 °C.
10. Resuspend the cells to 500 $OD_{600}$/ml and freeze for at least 1 h at −80 °C. Frozen spheroplasts can be stored at −80 °C for several weeks.
11. Thaw the spheroplasts in a room temperature water-bath, then add an equal volume of chilled, deionized water, and DTT and PMSF to 1 mM each.
12. Disrupt the cells with 10 strokes of the motor-driven homogenizer.
13. Sediment unbroken cells and large fragments by centrifugation at 2000 g for 5 min at 4 °C.
14. Collect the supernatant and sediment the membranes by centrifugation at 17 000 g for 15 min at 4 °C.
15. Resuspend the microsome pellet in B88 at 2500 OD/ml by gentle douncing in the 5 ml homogenizer.
16. Load 0.5 ml homogenate onto a 4 ml sucrose step gradient (2 ml each of 1.5 M and 1.2 M buffered sucrose solutions).
17. Sediment the membranes by centrifugation for 1.5 h at 200 000 g at 4 °C in the SW55 Ti rotor.
18. Aspirate the vacuole band on top of the 1.2 M sucrose step, then collect the microsomes at the 1.2 M/1.5 M sucrose interface with a P1000 Pipetman.
19. Wash the microsomes by dilution into 20 volumes B88, and sedimenting at 17 000 g for 15 min at 4 °C.
20. Resuspend the microsomes by gentle douncing in 1 ml B88. Measure the $OD_{280}$ of a 1:100 dilution in 2% SDS. Adjust the volume with B88 to a final concentration of $OD_{280}$ = 30.
21. Freeze the membranes in aliquots (20–200 μl as convenient) in liquid nitrogen and store them at −80 °C. Avoid freeze–thawing individual aliquots.

Protein translocation across the yeast ER membrane can be studied both co-translationally (see Protocol 4) and post-translationally (see Protocol 5), depending on the signal sequence of the translocation substrate (6). The most commonly used substrate is the post-translationally imported pheromone precursor prepro-α-factor (7–9).

## Protocol 4

## Co-translational protein translocation into yeast microsomes

### Equipment and reagents

- See Protocol 2
- Yeast microsomes (Protocol 3)
- Yeast translation extract (Protocol 1)
- mRNA for a secretory protein
- EDTA wash (RNase-free): 20 mM Hepes–KOH pH 7.4, 250 mM sucrose, 50 mM EDTA pH 8.0, 1 mM DTT
- Salt wash (RNase-free): 20 mM Hepes–KOH pH 7.4, 250 mM sucrose, 500 mM potassium acetate, 1 mM DTT
- Membrane storage buffer (RNase-free): 20 mM Hepes–KOH pH 7.4, 250 mM sorbitol, 50 mM potassium acetate, 1 mM DTT

### A. EDTA/high salt wash of microsomes

1. To 100 μl membranes add 100 μl EDTA wash. Incubate the membranes on ice for 15 min. Sediment the membranes by centrifugation in a refrigerated microcentrifuge at full speed for 10 min. Aspirate the supernatant.
2. Resuspend the membranes in 200 μl salt wash and incubate on ice for 1 h. Sediment the membranes as in step 1, then wash the membranes twice in 500 μl storage buffer.
3. Resuspend the membranes in the original volume (100 μl) storage buffer, aliquot, and freeze in liquid nitrogen.

### B. Co-translational protein translocation

1. Nuclease-treat an aliquot of yeast translation extract as described in Protocol 2.
2. Mix all components for *in vitro* translation except the mRNA and the DEPC-treated water in the ratios described in Protocol 2, step 3.
3. For individual translocation reactions, combine 7.5 μl translation mix (from step 2), 1 μg mRNA, DEPC-treated water to 10 μl, and 1 μl of EDTA/salt-washed microsomes.
4. Incubate at 20 °C for 1 h.

The active sites of both signal peptidase and oligosaccharyl transferase are located in the ER lumen, hence signal peptide cleavage and *N*-glycosylation can be used to monitor exposure of the respective sites in a substrate protein to the ER lumen (3, 7–9). Complete translocation across the ER membrane results in protection of the translocated protein from exogenously added proteases (7–9). Acquisition of protease-resistance can also reflect compact folding of a protein, therefore a control needs to be included in which the microsomes are solubilized by detergent prior to protease digestion of the sample, to ensure that the protein itself is not resistant to the protease used.

## Protocol 5

## Post-translational protein translocation into yeast microsomes

### Equipment and reagents

- Microcentrifuge
- SDS–PAGE equipment
- B88 (see Protocol 3)
- Yeast microsomes (see Protocol 3)
- *In vitro* translated prepro-a-factor (see Protocol 2)
- 10 × ATP stock: 10 mM ATP, 500 μM GDP-mannose, 400 mM creatine phosphate, 2 mg/ml creatine phosphokinase, in B88; frozen in 100 μl aliquots in liquid nitrogen, stored at −80 °C
- 20% (w/v) TCA
- Acetone
- SDS sample buffer
- Trypsin (Sigma): 1 mg/ml stock
- Proteinase K (Boehringer): 1 mg/ml stock
- 20% (w/v) Triton X-100
- 20% (w/v) SDS
- Con A buffer: 20 mM Tris–HCl pH 7.5, 500 mM NaCl, 1% Triton X-100
- Tris/NaCl buffer: 10 mM Tris–HCl pH 7.5, 50 mM NaCl
- 20% (v/v) Concanavalin A–Sepharose (Pharmacia) in Con A buffer

### A. Translocation

1. Thaw all reagents on ice. Mix B88 (to 20 μl final volume) and 2 μl of 10 × ATP, then add 1–2 μl yeast microsomes, $A_{280}$ = 30. Start the translocation by adding 2 μl (2–5 × $10^5$ cpm) *in vitro* translated prepro-α-factor.
2. Incubate the reaction at 20–30 °C for 15–60 min.
3. Add an equal volume of cold 20% TCA to each sample to stop the reaction. Incubate the samples on ice for 15–30 min. Sediment the proteins by centrifugation for 5 min at full speed in a microcentrifuge. Aspirate the supernatant.
4. Add 100 μl ice-cold acetone to each pellet, and centrifuge the samples for 5 min at full speed in a microcentrifuge. Aspirate the supernatant, air dry the pellets, and add 15–30 μl SDS sample buffer. Resuspend the samples by heating to 95 °C for 5 min. Analyse the proteins for translocation and glycosylation by SDS–PAGE on 12.5% gels.

### B. Protease protection

1. Translocate samples in triplicate. At the end of the translocation reaction, transfer the samples to ice.

**Protocol 5** continued

2 Within each set of three, add buffer only to the first sample, trypsin or proteinase K to a final concentration of 0.2 mg/ml to the second sample, and trypsin or proteinase K and Triton X-100 to a final concentration of 1% to the third sample.

3 Incubate the samples for 20 min on ice. Stop the digestion by TCA precipitation as above, and analyse the proteins for protease protection by SDS–PAGE.

### C. *N*-glycosylation

1 At the end of the translocation reaction, add SDS to 1%, and heat the samples to 95 °C for 5 min.

2 Add 1 ml Con A buffer and 50 μl of 20% Concanavalin A–Sepharose. Incubate the samples on a rotating wheel at room temperature for 2 h or in the cold room overnight.

3 Wash each sample three times with 1 ml Con A buffer, and once with 1 ml Tris/NaCl. Either take up the Sepharose in 30 μl SDS–PAGE sample buffer, heat to 95 °C for 5 min, and analyse the Concanavalin A-associated proteins by SDS–PAGE; or transfer the Concanavalin A–Sepharose to scintillation vials, add scintillation fluid, and count.

4 Alternatively, *N*-glycosylation can be analysed by immunoprecipitation of the translocated protein followed by digestion with endoglycosidase H (Boehringer) or protein *N*-glycanase (Boehringer), and analysing the resulting molecular weight shift by SDS–PAGE.

## 3 Interaction with chaperones and protein folding

During and after translocation into the ER lumen, secretory proteins have to acquire disulfide bonds, glycosyl side chains, and fold into an appropriate tertiary structure, before they can be packaged into ER-to-Golgi transport vesicles (10). These modifications and folding require multiple interactions with ER-resident enzymes and chaperones (10). Often, interactions need to take place in a specific order with specific chaperones, and with characteristic timing for individual substrate proteins (11). They can be monitored by photo- or chemical crosslinking of the translocated, radiolabelled substrate protein to ER-resident, unlabelled proteins (11). A method involving chemical crosslinking is described in Protocol 6. Subsequently, the crosslinked material can be immunoprecipitated with antibodies against either the folding substrate or the ER-resident protein. A large panel of homo- and heterobifunctional chemical crosslinkers is commercially available (Pierce), and the choice of crosslinker often is of critical importance for the outcome of the experiment. Features that should be taken into account when choosing a crosslinker include the abundance and distribution of specific amino acid side chains in the crosslinking partners, the size of the spacer between the two reactive groups in the crosslinker, and the buffer conditions in which the crosslinker will be used.

## Protocol 6

# Analysis of protein–protein interactions in the ER by chemical crosslinking

### Reagents

- Yeast wild-type or mutant microsomes
- *In vitro* translated, radiolabelled protein
- Dithiobis(succinimidylpropionate) (DSP) (Pierce)
- B88 (see Protocol 3), but pH 7.4!
- Dry dimethyl sulfoxide (DMSO) (Pierce)
- Ammonium acetate (Sigma)

### Method

1. Perform a 10 × scaled up translocation reaction (200 μl total volume, containing 20 μl microsomes of $A_{280}$ = 30). Chill the samples on ice for 2 min.
2. Sediment the membranes by centrifugation for 5 min at full speed in a refrigerated microcentrifuge.
3. Resuspend the microsomes in 1 ml of B88 pH 7.4, and repeat steps 2 and 3 twice.
4. Resuspend the final membrane pellet in 150 μl of B88 pH 7.4. Dissolve the DSP in DMSO at 5 mg/ml just prior to use. Add 6 μl DMSO containing 5 mg/ml DSP, or DMSO only as a control. Mix the samples by pipetting.
5. Incubate the samples at 24 °C for 20 min. The concentration of DSP and the length of the incubation time for crosslinking given here are reference points only; both should be titrated for individual protein– protein interactions.
6. Transfer the samples to ice and add 7.5 μl of 8.4 M ammonium acetate to quench the crosslinking reaction. Incubate the samples on ice for 20 min.
7. Add SDS to a final concentration of 1% to the samples, and heat the samples to 95 °C for 5 min (for crosslinking of soluble proteins) or 65 °C for 10 min (membrane proteins).
8. Immunoprecipitate the samples with saturating amounts of antibodies against the ER protein with which the translocated protein might interact (e.g. BiP). Analyse the immunoprecipitates by non-reducing SDS–PAGE to detect crosslinked material, or reducing SDS–PAGE to detect radiolabelled translocated protein associated with the non-radiolabelled, immunoprecipitated ER protein. Stable interactions can also be analysed without crosslinking by native immunoprecipitation from membranes solubilized at the end of the translocation reaction.

Secretory proteins which fail to fold in the ER lumen are retained in the ER (10). Many are subsequently exported across the ER membrane back to the cytosol and degraded by proteasomes, but some aggregate in the ER lumen instead (12, 13). Accumulation of misfolded proteins in the ER induces the Unfolded Protein Response (UPR), which, in intact cells, increases expression of ER-resident chaperones such as BiP and PDI (14). Translocation of a mutant protein that is unable to fold into microsomes generally results in increased amounts of complex formation with BiP and other chaperones, and in prolonged half-lives of these chaperone-folding substrate complexes (15).

## 4 Export of misfolded proteins from the ER to the cytosol for degradation

McCracken and Brodsky were the first to develop a cell-free assay for ER degradation of misfolded secretory proteins (16). The main reason for the long hiatus in developing such an assay despite repeated attempts over a period of almost a decade was the fact that ER degradation was thought to occur in the lumen of the ER (17). We now know that misfolded secretory proteins are transported retrogradely across the ER membrane, and are degraded by proteasomes in the cytosol (12). Cell-free assay systems based on mammalian microsomes or permeabilized cells have also been developed (18–20). The large variation in cytosol-dependence of cell-free assay systems for ER degradation is most likely a function of the number and activity of proteasomes associated with the microsomal preparations used in the respective assays.

In the yeast system, in order to achieve degradation of misfolded proteins *in vitro*, microsomes containing a misfolded, radiolabelled substrate need to be incubated in cytosol containing active proteasomes and ATP (16). The commonly used *in vitro* degradation substrate in yeast is a mutant derivative of the yeast pheromone precursor prepro-$\alpha$-factor in which all three *N*-glycosylation sites have been inactivated by site-directed mutagenesis and which, like its wild-type counterpart, translocates into microsomes post-translationally (16, 21). In the cases that have been investigated, mutants in genes required for misfolded protein export from the ER and degradation were also defective in the cell-free assay and caused increases in half-lives of misfolded proteins comparable to the increases observed in intact mutant cells (16, 22–25).

For some substrates, export to the cytosol and degradation by proteasomes is coupled, and chemical inhibition of proteasome activity with lactacystin or employing cytosol from a strain defective in proteasome function results in accumulation of these misfolded proteins in the microsomal lumen (reviewed in ref. 12). Other substrates are exported independently of the proteolytic activity of the proteasome; these proteins will accumulate in the yeast cytosol in a signal-cleaved, *N*-glycosylated form when proteasomes are inhibited (reviewed in ref. 12). In mammalian cytosol, proteins exported from the ER are rapidly deglycosylated by a cytosolic *N*-glycanase (26). This enzyme causes an asparagine to aspartate conversion at the deglycosylated site, which is indicative of a protein's passage through the ER followed by retrograde transport to the cytosol (26). A cytosolic *N*-glycanase has also been reported in *S. cerevisiae* (27), but its activity in cytosol derived from exponentially growing cells is too low to deglycosylate proteins exported from the ER within the time frame (1–2 h) of the cell-free assay described below (27). The preparation of yeast cytosol is described in Protocol 7, and a cell-free assay for monitoring the export and degradation of misfolded secretory proteins is described in Protocol 8.

## Protocol 7

## Preparation of yeast cytosol by liquid nitrogen lysis

### Equipment and reagents

- Electric blender with stainless steel vessel
- 10 litres YPD (see Protocol 1)
- Liquid nitrogen
- B88 (see Protocol 3)
- 1 M DTT
- 100 mM ATP

### Method

1. Grow up10 litres of cells in YPD to 4 $OD_{600}$/ml.
2. Harvest by centrifugation at 5000 g for 5 min. Wash once with 0.5 litres of deionized water, and once with 200 ml B88.
3. Resuspend the cells in about 5 ml B88 to the consistency of a thin paste.
4. Place a plastic beaker containing liquid nitrogen and a large stir bar onto a stir plate; pour cells into stirring liquid nitrogen in a thin stream; this results in small frozen cell pellets. Avoid forming large clumps. At this point, the cells can be kept frozen at −80 °C indefinitely.
5. In the cold room, pour liquid nitrogen into the pre-chilled blender, add the frozen cells, and blend continuously at high speed for 10 min to lyse the cells. During this period, keep adding nitrogen through the central hole in the lid to maintain the level just below the rim of the blender.
6. Pour the slurry into a beaker and allow the nitrogen to evaporate. This cell powder can be stored at −80 °C for several months.
7. Thaw the lysed cells in a 20 °C water-bath. Add DTT and ATP to 1 mM each. Sediment unbroken cells and large fragments by centrifugation for 10 min at 10 000 g and 4 °C.
8. Transfer the supernatant to ultracentrifuge tubes and sediment membranes at 100 000 g for 30 min. Transfer the supernatant to a new tube avoiding the lipid layer on top. At this point the cytosol can be frozen in aliquots in liquid nitrogen or gel filtered. The expected yield is 25–30 ml of 20–25 mg/ml protein concentration.

## Protocol 8

## Cell-free assay for misfolded secretory protein export from the ER and degradation

### Equipment and reagents

- B88 (see Protocol 3)
- 10 × ATP stock (see Protocol 4)
- 20% (w/v) TCA
- Liquid nitrogen lysed cytosol (Protocol 6) from a proteasome-proficient strain (12 mg/ml)

### Method

1. Perform an appropriately scaled up translocation reaction with mutant prepro-$\alpha$-factor precursor as described in Protocol 4. You will need the equivalent of a 20 μl translocation reaction per sample; volume of the translocation reaction = (number of samples + 1) × 20 μl.
2. Transfer the reaction to ice for 2 min, then sediment the membranes by centrifugation in a refrigerated microcentrifuge for 4 min at full speed.
3. Wash the membranes twice in 1 ml cold B88.
4. Resuspend the membranes in (number of samples + 1) × 10 μl of B88 containing 2 × ATP.
5. Add (number of samples + 1) × 10 μl cytosol (12 mg/ml) in B88, mix the samples by pipetting, and aliquot on ice into 20 μl aliquots.
6. Add 20 μl of 20% TCA to the t = 0 samples. Transfer all other samples to a 24 °C heating block; at each time point, transfer an aliquot to ice and add 20 μl 20% TCA.
7. Incubate samples on ice for at least 30 min, then process the samples for SDS–PAGE as described in Protocol 4A. Prepro-$\alpha$-factor and signal-cleaved pro-$\alpha$-factor are best resolved on 18% 4 M urea polyacrylamide gels, and can be quantified after overnight exposure to a phosphorimager screen.

## References


1. Walter, P. and Blobel, G. (1983). In *Methods in enzymology* (ed. S. Fleischer and B. Fleischer), Vol. 96, pp. 84–93. Academic Press, Inc.
2. Pelham, H. R. and Jackson, R. J. (1976). *Eur. J. Biochem.*, **67**, 247.
3. Blobel, G. and Dobberstein, B. (1975). *J. Cell Biol.*, **67**, 852.
4. Garcia, P. D., Hansen, W., and Walter, P. (1991). In *Methods in enzymology* (ed. C. Guthrie and G. R. Fink), Vol. 194, pp. 675–82. Academic Press, Inc.
5. Franzusoff, A., Rothblatt, J., and Schekman, R. (1991). In *Methods in enzymology* (ed. C. Guthrie and G. R. Fink), Vol. 194, pp. 662–74. Academic Press, Inc.
6. Ng, D. T. W., Brown, J. D., and Walter, P. (1996). *J. Cell Biol.*, **134**, 269.
7. Hansen, W., Garcia, P. D., and Walter, P. (1986). *Cell*, **45**, 397.
8. Rothblatt, J. A. and Meyer, D. I. (1986). *Cell*, **44**, 619.
9. Waters, M. G. and Blobel, G. (1986). *J. Cell Biol.*, **102**, 1543.

10. Hurtley, S. M. and Helenius, A. (1989). In *Annual reviews in cell biology* (ed. G. E. Palade, B. M. Alberts, and J. A. Spudich), Vol. 5, pp. 277–307. Annual Reviews Inc., Palo Alto, CA.
11. Tatu, U. and Helenius, A. (1997). *J. Cell Biol.*, **136**, 555.
12. Römisch, K. (1999). *J. Cell Sci.*, **112**, 4185.
13. Carrell, R. W. and Lomas, D. A. (1999). *Lancet*, **350**, 134.
14. Sidrauski, C., Chapman, R., and Walter, P. (1998). *Trends Cell Biol.*, **8**, 245.
15. Chillaron, J. and Haas, I. G. (2000). *Mol. Biol. Cell*, **11**, 217.
16. McCracken, A. A. and Brodsky, J. L. (1996). *J. Cell Biol.*, **132**, 291.
17. Klausner, R. D. and Sitia, R. (1990). *Cell*, **62**, 611.
18. Wilson, R., Allen, A. J., Oliver, J., Brookman, J. L., High, S., and Bulleid, N. J. (1995). *Biochem. J.*, **307**, 679.
19. Shamu, C. E., Story, C. M., Rapoport, T. A., and Ploegh, H. L. (1999). *J. Cell Biol.*, **147**, 45.
20. Xiong, X., Chong, E., and Skach, W. R. (1999). *J. Biol. Chem.*, **274**, 2616.
21. Mayinger, P. and Meyer, D. I. (1993). *EMBO J.*, **12**, 659.
22. Werner, E. D., Brodsky, J. L., and McCracken, A. A. (1996). *Proc. Natl. Acad. Sci. USA*, **93**, 13797.
23. Pilon, M., Schekman, R., and Römisch, K. (1997). *EMBO J.*, **16**, 4540.
24. Brodsky, J. L., Werner, E. D., Dubas, M. E., Goeckeler, J. L., Kruse, K. B., and McCracken, A. A. (1999). *J. Biol. Chem.*, **274**, 3453.
25. Gillece, P., Luz, J. M., Lennarz, W. J., de la Cruz, F. J., and Römisch, K. (1999). *J. Cell Biol.*, **147**, 1443.
26. Wiertz, E. J., Jones, T. R., Sun, L., Bogyo, M., Geuze, H. J., and Ploegh, H. L. (1996). *Cell*, **84**, 769.
27. Suzuki, T., Park, H., Kitajima, K., and Lennarz, W. J. (1998). *J. Biol. Chem.*, **273**, 21526.

# Chapter 8
# *In vitro* reconstitution of early to late endosome transport: biogenesis and subsequent fusion of transport intermediates

Feng Gu
Vollum Institute (L474), OHSU, Portland, OR 97201, USA.

Jean Gruenberg
Department of Biochemistry, Sciences II, University of Geneva, quai Ernest Ansermet 30, 1211 Geneva 4, Switzerland.

## 1 Introduction

Vesicles formed via invaginations of the plasma membrane are responsible for the internalization and turnover of cell surface proteins and lipids, and for the uptake of extracellular ligands and solutes. Although evidence is accumulating for the existence of more than one internalization pathway, most receptors are internalized via the well characterized clathrin-dependent pathway. The bulk of internalized molecules, including solutes, ligands, and membrane constituents, are then delivered to common early endosomes, at least in most animal cell types. From there, many receptors and lipids, as well as a significant proportion of internalized solutes, are rapidly recycled back to the extracellular medium, via recycling endosomes. In contrast, molecules destined to be degraded, including all down-regulated receptors, are selectively incorporated into transport intermediates destined for late endosomes. Protein sorting thus occurs within these common early endosomes, hence they have been referred to as sorting early endosomes (1). In this chapter, we will describe assays which have been used to study the sequential formation and consumption by fusion of transport intermediates between early and late endosomes.

Transport from early to late endosomes is mediated by relatively large vesicles (0.4–0.5 μm diameter) with a typical multivesicular appearance (2, 3), which will be referred to here as endosomal carrier vesicles/multivesicular bodies (ECV/MVBs). ECV/MVBs are formed on early endosomal membranes in a process that depends on some, but not all, components of the COP-I protein coat (4–7), which is also involved in the early secretory pathway (8). Membrane recruitment of endosomal COPs depends on the small GTP binding protein ARF1, like biosynthetic COPs and other coat complexes (9). However, ARF1 does not appear to mediate endosomal COP recruitment via activation of a phospholipase D or production of phosphatidic acid, in contrast to the mechanism proposed to operate in the biosynthetic pathway (10). Recent studies also indicate that βCOP, which is present on endosomes (4), interacts perhaps indirectly with a di-acidic signal (di-Glu) in the AIDS virus-encoded Nef protein, and that this interaction is responsible for CD4-Nef sorting in early endosome, and for its subsequent down-regulation (11). These studies support the view that early endosomal COPs facilitate the sorting of proteins destined for late endosomes and/or lysosomes. In addition, studies with ldlF cells, which contain a ts-mutation in the gene encoding for εCOP (12), have shown that COP inactivation at the restrictive temperature leads to a disruption of the early endosome ultrastructure, when ECV/MVB biogenesis is inhibited (6). Then, early endosome collapse into clusters of thin tubules, lacking the characteristic multivesicular domains corresponding to forming ECV/MVBs, which are normally observed on early endosomal membranes. These studies indicate that early endosome dynamics and/or organization is coupled to the biogenesis of ECV/MVBs.

In addition to endosomal COPs, the biogenesis of ECV/MVBs is also regulated by the acidic endosomal pH (13). Inhibition of the vacuolar ATPase causes both an inhibition of early to late endosome transport and disruption of the early endosome ultrastructure (13, 14) much like COP inactivation. In fact, this pH-dependent mechanism appears to be coupled biochemically and functionally to endosomal COP functions, since the association of both ARF1 and COPs to endosomal membranes is itself dependent on the acidic endosomal pH (9). This process differs from COP association to biosynthetic membranes, which is not pH dependent (4, 6). Endosomal and biosynthetic COPs also appear to differ in composition, since γ and δ COP are not present on endosomes (4, 6, 7). In addition, biosynthetic COPs interact with the KKXX endoplasmic reticulum retrieval motif (8), which is not present on endosomal proteins. Altogether, these differences point at the existence of somewhat plastic interactions between COP subunits and membrane constituents during coat recruitment and/or assembly, and suggest that COP functions may be differentially modulated on different sets of membranes. In the endocytic pathway, studies on the role of pH and COPs have lead to the proposal that pH changes in the endosomal lumen can be translated across the membrane by a pH-sensing mechanism, perhaps corresponding to pH-dependent conformational changes in endosomal transmembrane protein(s), and that this mechanism in turn regulates ARF1 and COP membrane association on the cytoplasmic face of early endosomal membranes (4, 6, 9). Since the pH of early endosomes is mildly acidic (~6.2) and then rapidly drops to 5.0–5.5 beyond early endosomes (15), whereas it raises in recycling endosomes at least in some cell types (15), it is attractive to believe that the pH-dependent mechanism of ECV/MVB biogenesis functions as a key regulatory step in the onset of the degradation pathway.

Once formed on early endosomal membranes, ECV/MVBs move towards late endosomes, which are often clustered in the perinuclear region in many cell types, and this movement depends on intact microtubules and motor proteins (2, 16, 17). Eventually, ECV/MVBs dock onto and fuse with late endosomes, in a process which depends on αSNAP, NSF, and perhaps another member of the triple A ATPase family (18), as well as presumably the small GTPase rab7 (19). In our studies, we designed an assay to reconstitute the formation of ECV/MVBs from early endosomes *in vitro*. ECV/MVBs formed *in vitro* exhibit the characteristic spherical and multivesicular appearance observed *in vivo*, and, in contrast to early endosomes, acquire the capacity to dock onto and fuse with late endosomes (4). We had also designed an *in vitro* assay to reconstitute docking/fusion of ECV/MVBs, either isolated from cells or generated *in vitro*, with late endosomes. This assay made it possible to investigate the role of microtubules and motor proteins in facilitating interactions between ECV/MVBs and late endosomes *in vitro*. A combination of these assays makes it possible to reconstitute sequentially the different steps involved in transport from early to late endosomes, including biogenesis, interactions with the cytoskeleton, and docking/fusion.

## 2 Baby hamster kidney (BHK21) cell culture

When setting up conditions with a new cell line, it is wise to use a freshly cloned cell line, and a limited number of passages (up to 20–30, depending on the cell type), in order to reduce variations from experiment to experiment. The *in vitro* assays described in this chapter utilize endosomal fractions prepared from cloned BHK cells. Optimal conditions for cell homogenization and fractionation, hence for the transport assays, depend on optimal culture conditions. Monolayers of BHK21 cell line are grown and maintained in Glasgow's minimum essential medium (G-MEM, Sigma cell culture reagents) supplemented with 5% fetal calf serum (FCS, Sera-Technologies), 10% tryptose phosphate broth (Gibco BRL), 1% glutamine (2 mM, Gibco BRL), 1% penicillin–streptomycin (100 IU/ml–100 UG/ml, Gibco BRL) at 37 °C in a 5% $CO_2$ incubator. Cells are maintained in culture by two passages each week of 1:20 dilutions by surface area. For experiments, cells are always seeded after 1:4 dilution by surface area from a confluent 10 cm Petri dish of three days old cells at a density of $\sim 4 \times 10^4$ cells/cm$^2$ of culture dish. Cells are allowed to attach and grow to confluency for 14–16 h. With this protocol, cells form an even monolayer on the dish (corresponding to $\sim 1.5 \times 10^7$ cells). This is important to ensure that markers can be evenly internalized within all cells, and that cells are easily homogenized with minimal damage to endosomes. Typically, one may expect to obtain from a 10 cm Petri dish, 50–100 μg of early endosomal fraction and 10–20 μg of late endosomal fraction. For large scale preparation of endosomes, four square (20 × 20 cm, Nunc) dishes are seeded with BHK cells from ten confluent 10 cm Petri dishes, yielding ~3 mg of early endosomal and ~0.5 mg of late endosomal fractions.

## 3 Labelling of the different endosomal compartments in BHK cells

Early and late endosomes in BHK cells can be identified morphologically and biochemically using markers of these compartments, e.g. the small GTPase rab5 (20) and EEA1 (21) in early endosomes, or rab7 (20) and lyso-bisphosphatidic acid in late endosomes (22). Endosomes can also be labelled with fluid phase markers, or other endocytosed tracers, providing the additional advantage that the same marker can often be positioned selectively in different endosomal populations, depending on the incubation conditions (time, temperature, drugs). In transport assays, endosomes are conveniently labelled after fluid phase endocytosis of the desired tracer, as follows:

(a) Early endosome: cells are incubated for 5–10 min at 37 °C in the presence of the tracer.
(b) ECV/MVB: cells are incubated after depolymerization of the microtubule network for 10–15 min at 37 °C in the presence of the tracer, followed by a 30–35 min incubation in tracer-free medium.
(c) Late endosomes: conditions are identical to ECV/MVB labelling, except that the microtubule network is intact (23).

When monitoring ECV/MVB biogenesis, we use horseradish peroxidase (HRP) as a fluid phase tracer to label the donor early endosomes, because the enzyme exhibits a very high specific activity, is easily quantified biochemically, and can be revealed morphologically (by light and electron microscopy). In the docking/fusion assay, we use biotinylated HRP (bHRP) and avidin as markers. These are endocytosed separately into two cell populations (23), and, after fractionation, endosomes labelled with each marker are mixed in the assay. If fusion occurs, a product is specifically formed, which can be immunoprecipitated with anti-avidin antibodies, and then the enzymatic activity of bHRP can be quantified. When ECV/MVBs are first generated *in vitro* and then their docking/fusion capacity is measured, bHRP, instead of HRP, is used to label donor early endosomes (4). At the next step, the fusion activity of bHRP-labelled, ECV/MVBs generated *in vitro* is measured using acceptor endosomes labelled with endocytosed avidin. Protocols 1 and 2 describe typical conditions used to label early endosomes, used as donor membranes in the assay measuring ECV/MVB biogenesis, and late endosomes, used as acceptor membranes for ECV/MVBs generated *in vitro*. Obviously, conditions of internalization can be changed to accommodate other experimental conditions.

All experimental steps and all solutions are kept at ice temperature except internalization medium which is pre-warmed to 37 °C. Throughout the protocol, care must be taken to limit damage to the monolayer, and to ensure that cells do not dry. The conditions in Protocol 1 are described for one square dish of 20 × 20 cm. HRP biotinylation is described in ref. 2. When the assay is used strictly to follow ECV/MVB biogenesis, and not subsequent fusion, bHRP is replaced with HRP (8 mg/ml).

## Protocol 1

### *In vivo* labelling of donor early endosomes with bHRP (for *in vitro* biogenesis of ECV/MVBs)

#### Equipment and reagents

- Water-bath at 37 °C
- Ice plate: a metal plate is fitted into a flat ice bucket to guarantee good contact with the dish
- Rocking platform
- Phosphate-buffered saline (PBS): 137 mM NaCl, 2.7 mM KCl, 1.5 mM $KH_2PO_4$, 8.1 mM $Na_2HPO_4$
- Biotinylated HRP (bHRP)
- Internalization medium (IM):10 mM Hepes pH 7.4, 10 mM D-glucose in MEM
- IM-bHRP: internalization medium containing 1.8 mg/ml bHRP, freshly prepared and filtered through a 0.45 mm filter
- PBS-BSA: PBS containing 5 mg/ml BSA (IgG-free BSA is not necessary)

#### Method

1. Place dishes of cells on an ice plate (a metal plate is fitted into a flat ice bucket to guarantee good contact with the dish).
2. Remove medium with an aspirator connected to a water pump, and rinse the cell monolayer three times with 15 ml PBS. PBS is added carefully from the side of the dish, to limit damage to the monolayer.
3. Aspirate PBS, transfer the dish to a water-bath at 37 °C, on a flat metal plate with the bottom of the dish in contact with the water.
4. Rapidly, but carefully, pour 25 ml of internalization medium containing 1.8 mg/ml bHRP (IM-bHRP) on the dish and incubate for 10 min at 37 °C.
5. Rapidly remove the dish, place it back onto the ice plate, aspirate the warm IM-bHRP, and carefully pour 15 ml of ice-cold PBS-BSA onto the monolayer.
6. Place the ice bucket onto a rocking platform and gently rock for 10 min, to ensure proper washing of the cell monolayer. The solution is then aspirated and replaced with fresh PBS-BSA, and the washing step is repeated twice (three times in total).
7. Aspirate PBS-BSA and rinse twice with PBS. The cells are now ready for homogenization (see Protocol 3).

## Protocol 2

### *In vivo* labelling of acceptor late endosomes with avidin (for ECV/MVB docking/fusion assay)

#### Equipment and reagents

- See Protocol 1
- IM-BSA: internalization medium containing 2 mg/ml BSA
- IM-avidin: internalization medium (see Protocol 1) containing 3.2 mg/ml avidin, freshly prepared and filtered through a 0.45 μm filter

#### Method

1. Avidin is internalized into living BHK cells as described in Protocol 1, steps 1–4, except that IM-bHRP is replaced with IM-avidin.
2. Rapidly remove the dish, place it back onto the ice plate, aspirate the warm IM-avidin, and carefully pour 15 ml of ice-cold PBS-BSA onto the monolayer.
3. Place the ice bucket onto a rocking platform and gently rock for 5 min. The solution is then aspirated and replaced with fresh PBS-BSA, and the washing step is repeated once (twice in total).
4. Aspirate the ice-cold PBS-BSA, and place the dish back in the 37 °C water-bath. Pour on 40 ml of IM-BSA pre-warmed to 37 °C, and incubate for 40 min at 37 °C.
5. Return the dish onto the ice plate, rinse twice with ice-cold PBS. The cells are now ready for homogenization.

## 4 Homogenization and subcellular fractionation of BHK cells

In the assays described here, early or late endosomal fractions are prepared by subcellular fractionation and then used *in vitro*. The fractionation protocol was established in BHK cells, and is based on endosome flotation in a simple step gradient (24). Conditions may need to be changed when using different cell types. After fractionation on this gradient, membranes are separated into three major fractions that are selectively enriched in:

(a) The bulk of biosynthetic membranes and the plasma membrane.

(b) Early endosomes and intermediate compartment/*cis*-Golgi network.

(c) Late endosomes as well as ECV/MVBs present at steady-state.

In some experiments, cells are treated with Brefeldin A prior to fractionation, in order to deplete biosynthetic membranes from early endosomal fractions (25). Using these fractions, we have shown that both early and late endosomes exhibit homotypic fusion activity, but that early and late endosomes do not directly fuse with each other. We have also shown that ECV/MVBs, generated *in vivo* or *in vitro*, acquire the capacity to fuse with late endosomes.

Cells labelled with the desired endocytosed tracer (bHRP or avidin) are first homogenized through a 22G needle fitted onto a 1 ml Tuberculin syringe. Gentle conditions of homogenization should be used to limit damage to endosomes, and loss of the fluid phase markers. Homogenization can be easily followed using phase contrast microscopy, by following the release of intact and clean nuclei upon cell breakage. However, a more precise measurement of homogenization is necessary. Indeed, endosome latency (percentage of marker remaining intravesicular after homogenization) should remain >70% after homogenization for the assays described here. Latency can be easily measured after high speed centrifugation of the post-nuclear supernatant, as described in ref. 26. In addition, the fractionation described in Protocol 3, as for any other fractionation protocol, should be analysed in detail, and a balance sheet (enrichment and yield) should be established as described in ref. 26, to ensure that experimental conditions are optimal.

Conditions in Protocol 3 are described for a square dish of 20 × 20 cm. It is often convenient to work in a cold room, to ensure that cells and cell extracts remain at 4 °C. Precise sucrose content of the solutions used in the gradient must be measured using a refractometer.

## Protocol 3

## Fractionation of tissue culture cells

### Equipment and reagents

- SW60 or SW40 tubes and rotors
- Peristaltic pump for collecting fractions after centrifugation
- 1 ml Tuberculin syringe fitted with a 22G needle
- PBS (see Protocol 1)
- Homogenization buffer (HB): 250 mM sucrose, 3 mM imidazole pH 7.4
- Sucrose solutions: 62%, 35%, and 25% sucrose (w/w) in 3 mM imidazole pH 7.4

### Method

1. Once markers have been internalized into endosomes (see Protocols 1 and 2), aspirate the solution, and pour 12 ml of ice-cold PBS onto the monolayer which must remain in the ice-bath at all steps. Scrape the cells off the dish using a rubber policeman (with the edge cut off at a low angle to limit damage to the cells) at 4 °C.
2. Collect the cells in a centrifuge tube and sediment at 1000 *g* for 5 min at 4 °C.
3. Resuspend cells very gently in 3 ml HB using a plastic Pasteur pipette with a wide tip (or a blue tip which has been cut off), and re-centrifuge at 1500 *g* for 10 min at 4 °C.

**Protocol 3** continued

4 Remove the supernatant, add 150 μl of HB containing a cocktail of protease inhibitors (10 μM leupeptin, 10 μg/ml aprotinin, 1 μM pepstatin) to a 100 μl cell pellet. Resuspend the cell pellet with a blue tip, and then pass the cells through a 22G needle fitted onto a 1 ml Tuberculin syringe. Monitor by phase contrast microscopy. Nuclei should be clean of cellular materials and intact.

5 Centrifuge the homogenate at 1500 g for 10 min at 4 °C and collect the post-nuclear supernatant (PNS).

6 Adjust the PNS to 40.6% sucrose using a stock solution of 62% sucrose (w/w) in 3 mM imidazole pH 7.4 (~1:1.2 dilution of PNS).

7 Load 1 ml of the PNS in 40.6% sucrose at the bottom of an SW60 centrifuge tube. Overlay sequentially with 1.5 ml of 35% sucrose (w/w) in 3 mM imidazole pH 7.4, and then with 1 ml of 25% sucrose (w/w) in 3 mM imidazole pH 7.4, and eventually fill the tube with HB. When necessary, for large scale preparations, the PNS obtained from four square dishes (20 × 20 cm) is adjusted to 40.6% sucrose and loaded in six SW40 tubes, each being sequentially filled with 4.5 ml of 35% sucrose, 3 ml of 25% sucrose, and HB.

8 Mount the tube in the appropriate cold rotor (SW60 or SW40), and centrifuge at 4 °C for 60 min (SW40) or 90 min (SW60) at 35 000 rpm.

9 After centrifugation, collect early and late endosomal fractions from the 35%/25% sucrose interface and the 25%/HB interface, respectively, using a peristaltic pump fitted with a 50 μl glass capillary tube. Alternatively, a 200 μl tip can be used to collect the fractions manually.

10 Fractions are now ready for the transport assays. Fractions can also be aliquoted, flash frozen in liquid nitrogen, and stored at −80 °C.

## 5 *In vitro* reconstitution of the ECV/MVB formation from early endosomes

Early endosomes labelled with HRP are used as donor fractions for generating ECV/MVBs *in vitro*. This method is described in Protocol 4. Since the amount of ECV/MVBs generated *in vitro* is relatively low, large amount of donor early endosomes (250–300 μg) are used for each point in the assay, hence large scale preparations are useful. Early endosomes are mixed with rat liver cytosol (the protocol for rat liver cytosol preparation is described in Protocol 6), in the presence of an ATP regenerating system at 37 °C for 30 min to allow formation of ECV/MVBs *in vitro*. Then, the reaction mixture is adjusted to 25% sucrose, loaded at the bottom of an SW60 centrifuge tube, and overlaid with HB. Vesicles formed *in vitro* are separated from the donor early endosome membranes by flotation in a second sucrose step gradient. After centrifugation, the ECV/MVBs generated *in vitro* are collected from the 25%/HB interface, whereas donor early endosomes sediment under these conditions

The budding efficiency is calculated as the ratio of HRP incorporated into ECV/MVBs over the total amount of HRP present (donor membranes and ECV/MVBs formed *in vitro*), and expressed as a percentage. The protocol used for measuring HRP activity is described in Protocol 7. Efficiency normally corresponds to ~10%, a value that reflects the early endosomal volume entrapped during one round of vesicle formation *in vitro*, and which agrees well with *in vivo* measurements. ECV/MVB formation is dependent on ATP, cytosolic proteins, and does not occur at 4 °C. Biogenesis of ECV/MVBs is inhibited after neutralization of the early endosomal pH *in vitro*, much like *in vivo*. Hence, it is important to use $Cl^-$ as counter anion in the solutions, since $Cl^-$ contributes to proper maintenance of the endosomal pH (27). In this assay, as in all transport assays, care must be taken to minimize osmotic, chemical, and oxidative stress.

## Protocol 4

## *In vitro* biogenesis of ECV/MVBs

### Equipment and reagents

- Water-bath at 37 °C
- 50 × concentrated buffer: 1 M Hepes pH 7.2, 1 M $MgOAc_2$, 1 M DTT
- Homogenization buffer (HB) (see Protocol 3)
- Rat liver cytosol (see Protocol 6)
- Early endosomal fraction (see Protocols 1 and 3)
- ATP regenerating system (prepare at 4 °C just before the experiment by mixing equal volumes of the following stock solutions): 100 mM ATP pH 7.0 (Sigma), 800 mM creatine phosphate (Boehringer), 4 mg/ml creatine kinase (800 U/ml, Boehringer) in 50% glycerol

### Method

1. For one point in the assay, mix in the following order in a 15 ml Falcon tube:
   - 27 μl of 50 × concentrated buffer
   - 105 μl of 1 M KCl
   - 300 μl of HB
   - 300 μl of rat liver cytosol (see Protocol 6)
   - 300 μl of early endosomal fraction (labelled with bHRP or HRP)
   - 75 μl of ATP regenerating system

   If needed, the ATP regenerating system can be replaced with an ATP-depleting system, using 20 μl apyrase (1200 U/ml, Sigma).
2. Mix gently and incubate for 5 min on ice (to reduce osmotic stress).
3. Incubate the reaction mixture at 37 °C for 15–30 min (using a water-bath).
4. Bring the mixture back to ice, and cool to ice temperature. The sample is then ready for the separation of vesicles formed *in vitro* from donor membranes (see Protocols 5a and 5b).

If ECV/MVBs generated *in vitro* are to be used for quantification of the process, HRP is used to label early endosomes *in vivo*.

## Protocol 5a

## Separation of ECV/MVBs formed *in vitro* from donor early endosomal membranes (direct measurement of ECV/MVB biogenesis)

### Equipment and reagents

- See Protocol 3
- Beckman TL-100 ultracentrifuge
- TLS55 tubes and rotor
- HB containing 0.2% Triton X-100

### Method

1. Adjust the reaction mixture prepared in Protocol 3 to 25% sucrose, using a 62% sucrose stock solution in 3 mM imidazole pH 7.4, and bring to a final volume of 2 ml with 25% sucrose in 3 mM imidazole pH 7.4.
2. Load the reaction mixture (adjusted to 25% sucrose) at the bottom of an SW60 centrifuge tube and overlay with HB to fill the gradient. If fusion activity of donor membranes needs to be measured after the assay, it is wiser to load the reaction mixture on top of a 50% sucrose cushion in 3 mM imidazole pH 7.4, to avoid membrane damage during centrifugation.
3. Centrifuge at 35 000 rpm for 60 min at 4 °C.
4. Carefully collect the fraction containing ECV/MVBs generated *in vitro* (~300 μl), which appears as a faint band at the interface of 25% sucrose/HB, using a peristaltic pump (or a 200 μl pipette tip). Then, dilute this fraction with 1 ml of HB, and re-centrifuge for 30 min in a TLS55 rotor (using a table-top Beckman TL-100 ultracentrifuge) at 55 000 rpm to sediment vesicles. This step allows concentration of ECV/MVBs and their separation from free HRP which may have been released by vesicle breakage during the assay. Dissolve the pellet containing the vesicles formed *in vitro* in 500 μl of HB plus 0.2% Triton X-100. Use 50 μl to measure HRP activity.
5. Collect the pellet containing the donor membranes at the bottom of the tube, and resuspend this pellet in 1 ml of HB containing 0.2% Triton X-100. Use 25 μl to measure HRP activity (this fraction can be used for SDS gel electrophoresis and Western blotting, if needed). If the fusion activity of donor membranes needs to be tested, the fraction is collected, Triton X-100 being omitted, and adjusted to conditions used to measure docking fusion (see Protocol 8).
6. The total HRP activity is calculated from the sum of HRP present in the fractions containing vesicles formed *in vitro* and the donor membranes, and budding efficiency is expressed as a percentage of the total HRP present.

If ECV/MVBs generated *in vitro* are to be used in the cell-free fusion assay, bHRP is used to label early endosomes *in vivo*.

## Protocol 5b

### Separation of ECV/MVBs formed *in vitro* from donor early endosomal membranes (for subsequent use in docking/fusion assay)

#### Equipment and reagents

- See Protocol 5a

#### Method

1. Follow Protocol 5a, steps 1–3. The reaction mixture is loaded on top of a 50% sucrose cushion in 3 mM imidazole pH 7.4, to avoid membrane damage during centrifugation.
2. Follow Protocol 5a, step 4. ECV/MVBs generated *in vitro* are diluted and then loaded on top of a 50% sucrose cushion in 3 mM imidazole pH 7.4 to limit damage during centrifugation. Vesicles formed *in vitro* are collected with a peristaltic pump from the 50%/buffer interface. They are not dissolved in Triton X-100, but adjusted to conditions used in the fusion assay (as described in Protocol 8). Keep a 25–50 μl aliquot to measure HRP activity.
3. If needed, donor membranes are collected with a peristaltic pump at the 50%/25% interface using a peristaltic pump, and adjusted to conditions used in the fusion assay (as described in Protocol 8). Keep a 25–50 μl aliquot to measure HRP.

Rat liver cytosol is a convenient source of large amounts of cytosol. Cytosol from BHK cells, or other sources, can also be used (26).

## Protocol 6

### Rat liver cytosol preparation

#### Equipment and reagents

- Sorvall SS-34 tubes and rotor
- SW40 tubes and rotor
- Homogenization buffer (HB) (see Protocol 3)

#### Method

1. Two livers are removed from rats, washed in HB, weighed, and homogenized using an electrical mixer in a volume (ml) of HB corresponding to 5 × weight of livers (g), in the presence of a cocktail of protease inhibitors (10 μM leupeptin, 10 μg/ml aprotinin, 1 μM pepstatin, 1 mM PMSF, 1 mM benzamidine) at 4 °C (9, 28).

**Protocol 6** continued

2 The mixture is centrifuged at 5000 rpm in a Sorvall SS-34 rotor for 10 min at 4 °C.

3 The supernatant is centrifuged at 15 000 rpm for 10 min in the same rotor at 4 °C.

4 The supernatant is centrifuged at 35 000 rpm for 60 min in an SW40 rotor.

5 The supernatant is aliquoted, flash-frozen in liquid nitrogen, and stored in liquid nitrogen.

The substrates of the peroxidase are *o*-dianisidine and $H_2O_2$. The brown coloured product is quantified in a spectrophotometer at 455 nm.

## Protocol 7

## Determination of HRP enzymatic activity

### Equipment and reagents

- Sterilized glassware, in order to limit contamination of the mixture
- Homogenization buffer (HB) (see Protocol 3)
- HRP reaction mixture: mix 12 ml of 0.5 M Na phosphate buffer pH 5.0, 6 ml of 2% Triton X-100, and 100.8 ml of bidistilled water. Add 13 mg *o*-dianisidine, dissolve gently, and add 1.2 ml of 0.3% $H_2O_2$. The reagent can be kept at 4 °C in the dark for one to two weeks, as long as a straw colour does not develop.

### Method

1 Prepare standards containing known amounts of HRP in 200 μl of HB (10 ng to 1 μg, depending on the time of the reaction). Samples and blanks are prepared in the same manner.

2 Add 1 ml of the HRP reaction mixture to each tube, and record the time.

3 When a yellow colour develops, stop the reaction with 20 μl of 10% $NaN_3$ and read the absorbance at 455 nm. HRP activity is expressed as OD (optical density) units/min.

## 6 *In vitro* fusion of the ECV/MVBs with late endosomes

ECV/MVBs formed *in vitro* acquire the competence to undergo fusion with late endosomes, much like ECV/MVBs isolated from cells. Thus, they are fully competent to function as transport intermediates between early and late endosomes. In contrast, early endosomes are not fusogenic with late endosomes *in vitro*, yet both endosomal populations exhibit homotypic fusion activity. Previous studies showed that molecular requirements of early endosome homotypic fusion differ from those necessary to generate ECV/MVBs. Whereas early endosome fusion depends on the small GTPase rab5 (24), ECV/MVB formation is rab5-independent (4). Conversely, ECV/MVB formation, but not early endosome fusion, requires an active endosomal COP complex and depends on an acidic endosomal pH (4, 6). The endosome fusion assays have been described in detail (4, 16, 17, 23, 24, 29). As mentioned above, the assay is based on the separate internalization of the fusion markers (bHRP and avidin) into two cell populations. In order to study the fusion capacity of ECV/MVBs generated *in vitro*, ECV/MVBs are prepared from donor early endosomal membranes labelled with bHRP. In parallel, acceptor late endosomes labelled with avidin are prepared by fractionation. In the assay, bHRP-labelled ECV/MVBs are mixed with avidin-labelled late endosomes. Whereas ECV/MVB–late endosomes interactions are facilitated by the presence of polymerized microtubules in the assay reactions, docking/fusion is in itself microtubule-independent (4, 16), and thus polymerized microtubules are not necessary when fusion capacity is being recorded. At the end of the reaction, the bHRP–avidin product formed in the endosomal lumen upon fusion, is then immunoprecipitated with anti-avidin antibodies, and the enzymatic activity of bHRP is quantified. The fusion assay is described in Protocol 8, and the method for detecting fusion is described in Protocol 9.

## Protocol 8

## *In vitro* fusion assay

### Reagents

- See Protocol 4
- bHRP-labelled ECV/MVBs
- Avidin-labelled late endosomes
- Rat liver cytosol (see Protocol 6)
- 30 μg/ml biotinylated insulin in PBS
- 20% Triton X-100 in PBS
- PBS-BSA: PBS containing 5 mg/ml BSA

### Method

1 For one detection point, mix the following solutions in the same order as indicated below, at the bottom of an Eppendorf tube:
   - 3 μl of 50 × concentrated buffer (Protocol 4)
   - 11 μl of 1 M KOAc (or 1 M KCl)
   - 8 μl of 30 μg/ml biotinylated insulin in PBS (to quench free avidin which may be released during the reaction, but biotinylated insulin must be omitted when measuring the total avidin–bHRP complex which can be formed in the presence of detergent)
   - 10 μl of the ATP regenerating system (or 8 μl of apyrase, when ATP needs to be depleted)
   - 50 μl of bHRP-labelled ECV/MVBs formed *in vitro* (~5–10 μg)
   - 50 μl of purified avidin-labelled late endosomes
   - 30 μl of rat liver cytosol

**Protocol 8** continued

In some experiments, the avidin-labelled late endosome fraction can be replaced by 70 μl of a post-nuclear supernatant prepared from cells containing avidin-labelled late endosomes. The mixture is then kept on ice for 3–5 min, for buffers and salts to equilibrate, in order to reduce osmotic stress.

2. Mix gently and incubate at 37 °C for 45 min in a water-bath. Do not agitate during the incubation.
3. Place the reaction mixture in an ice-bath, and cool to ice temperature for 2–3 min.
4. Add 5 μl of 30 μg/ml biotinylated insulin in PBS, to quench the avidin which may not have reacted during fusion. When measuring the total complex which can be formed in detergent, biotinylated insulin is omitted (if needed, 10 μl of 0.17 mg/ml bHRP can then be added).
5. Add 7.5 μl of 20% Triton X-100 in PBS, incubate for 30 min on ice, and dilute to 0.9 ml with PBS-BSA.
6. Centrifuge at 5000 rpm for 3 min in an Eppendorf centrifuge, and collect the supernatant.
7. Quantify the avidin–bHRP formed upon fusion and present in the supernatant as described below.

## Protocol 9

## Detection of fusion

### Reagents

- Antibodies against avidin (obtained from various commercial sources)
- Protein A–Sepharose
- PBS-BSA: PBS containing 5 mg/ml BSA
- PBS-BSA containing 0.2% Triton X-100

### Method

1. For immunoprecipitation, use 50 μl of the commercial protein A–Sepharose slurry per point. Wash the beads by resuspending 50 μl protein A–Sepharose slurry in 1 ml PBS-BSA, and sediment the beads in an Eppendorf centrifuge for 2–3 min. Repeat this step five times.
2. Resuspend the slurry to a final volume of 1 ml with PBS-BSA and add 100 μg of affinity-purified anti-avidin antibody for 1 h at 4 °C.
3. Wash the beads five times in PBS-BSA and resuspend in 100 μl of PBS-BSA.
4. Add 100 μl of anti-avidin antibody-coupled protein A–Sepharose beads to the supernatant obtained from Protocol 8, step 7, and rotate end-over-end for at least 1 h at 4 °C.

**Protocol 9** continued

5. Wash the beads five times with PBS-BSA containing 0.2% Triton X-100 and once with PBS.
6. Resuspend the beads in 100 μl PBS and assay for HRP as described in Protocol 7. When the beads become yellow, vortex the beads, sediment in an Eppendorf centrifuge, and use the supernatant for optical density (OD) measurement at 455 nm.
7. The efficiency of the reaction is expressed as a ratio of the HRP activity detected in the assay over the total amount of activity measured in the presence of detergent.

## 7 Conclusion

The assays described above have been used to identify and characterize some of the molecular requirements of sequential steps of transport from early to late endosomes, including vesicle formation and docking/fusion. These assays will be useful to characterize some of the new factors which are currently being identified. In addition, these assays make it possible to efficiently dissect the molecular requirements at each step in this pathway. We are currently using this strategy to gain new insights into the mechanisms regulating membrane organization, dynamics, and remodelling along the degradation pathway of animal cells.

## References

1. Mayor, S., Presley, J. F., and Maxfield, F. R. (1993). *J. Cell Biol.*, **121** (6), 1257.
2. Gruenberg, J., Griffiths, G., and Howell, K. E. (1989). *J. Cell Biol.*, **108** (4), 1301.
3. Gruenberg, J. and Maxfield, F. R. (1995). *Curr. Opin. Cell Biol.*, **7** (4), 552.
4. Aniento, F., Gu, F., Parton, R. G., and Gruenberg, J. (1996). *J. Cell Biol.*, **133** (1), 29.
5. Daro, E., Sheff, D., Gomez, M., Kreis, T., and Mellman, I. (1997). *J. Cell Biol.*, **139** (7), 1747.
6. Gu, F., Aniento, F., Parton, R. G., and Gruenberg, J. (1997). *J. Cell Biol.*, **139** (5), 1183.
7. Whitney, J. A., Gomez, M., Sheff, D., Kreis, T. E., and Mellman, I. (1995). *Cell*, **83** (5), 703.
8. Lowe, M. and Kreis, T. E. (1998). *Biochim. Biophys. Acta*, **1404** (1–2), 53.
9. Gu, F. and Gruenberg, J. (2001). *J. Biol. Chem.*, **275**, 8154.
10. Ktistakis, N. T., Brown, H. A., Waters, M. G., Sternweis, P. C., and Roth, M. G. (1996). *J. Cell Biol.*, **134** (2), 295.
11. Piguet, V., Gu, F., Foti, M., Demaurex, N., Gruenberg, J., Carpentier, J. L., *et al.* (1999). *Cell*, **97** (1), 63.
12. Guo, Q., Penman, M., Trigatti, B. L., and Krieger, M. (1996). *J. Biol. Chem.*, **271** (19), 11191.
13. Clague, M. J., Urbe, S., Aniento, F., and Gruenberg, J. (1994). *J. Biol. Chem.*, **269** (1), 21.
14. Bayer, N., Schober, D., Prchla, E., Murphy, R. F., Blaas, D., and Fuchs, R. (1998). *J. Virol.*, **72** (12), 9645.
15. Yamashiro, D. J. and Maxfield, F. R. (1984). *J. Cell. Biochem.*, **26** (4), 231.
16. Aniento, F., Emans, N., Griffiths, G., and Gruenberg, J. (1993). *J. Cell Biol.*, **123** (6 Pt 1), 1373.

17. Bomsel, M., Parton, R., Kuznetsov, S. A., Schroer, T. A., and Gruenberg, J. (1990). *Cell*, **62** (4), 719.
18. Robinson, L. J., Aniento, F., and Gruenberg, J. (1997). *J. Cell Sci.*, **110** (Pt 17), 2079.
19. Feng, Y., Press, B., and Wandinger-Ness, A. (1995). *J. Cell Biol.*, **131** (6 Pt 1), 1435.
20. Chavrier, P., Parton, R. G., Hauri, H. P., Simons, K., and Zerial, M. (1990). *Cell*, **62** (2), 317.
21. Mu, F. T., Callaghan, J. M., Steele-Mortimer, O., Stenmark, H., Parton, R. G., Campbell, P. L., *et al.* (1995). *J. Biol. Chem.*, **270** (22), 13503.
22. Kobayashi, T., Stang, E., Fang, K. S., de Moerloose, P., Parton, R. G., and Gruenberg, J. (1998). *Nature*, **392** (6672), 193.
23. Gruenberg, J. and Howell, K. E. (1989). *Annu. Rev. Cell Biol.*, **5**, 453.
24. Gorvel, J. P., Chavrier, P., Zerial, M., and Gruenberg, J. (1991). *Cell*, **64** (5), 915.
25. Rojo, M., Pepperkok, R., Emery, G., Kellner, R., Stang, E., Parton, R. G., *et al.* (1997). *J. Cell Biol.*, **139** (5), 1119.
26. Gruenberg, J. and Gorvel, J. P. (1992). In *Protein targeting: a practical approach* (ed. A. I. Magee and T. Wileman), pp. 187–216. Oxford University Press, Oxford.
27. Mellman, I., Fuchs, R., and Helenius, A. (1986). *Annu. Rev. Biochem.*, **55**, 663.
28. Aniento, F., Roche, E., Cuervo, A. M., and Knecht, E. (1993). *J. Biol. Chem.*, **268** (14), 10463.
29. Gruenberg, J. E. and Howell, K. E. (1986). *EMBO J.*, **5** (12), 3091.

# Chapter 9
# Receptor biology

Mercedes Dosil*
Centro de Investigacion del Cancer, Universidad de Salamanca, Salamanca, Spain.

Pamela Mentesana and James B. Konopka
Department of Molecular Genetics and Microbiology, State University of New York, Stony Brook, Stony Brook, NY 11794-5222, USA.

## 1 Introduction

Cell surface receptors for hormones, neurotransmitters, and other ligands have long been recognized to play a key role in cellular regulation. Thus, it is with great interest that genome sequencing projects are revealing the existence of new receptors that have not been studied previously. Therefore, we will describe protocols in this chapter that will be helpful for the fundamental characterization of new receptors. The methods will emphasize G protein-coupled receptors since this receptor family is expected to grow by hundreds of new members. However, the techniques will be generally applicable to a wide range of receptors and will also be useful for investigators wishing to add new approaches to previously characterized receptors.

## 2 Ligand binding assays

### 2.1 General considerations

This section will describe two protocols that will serve as general examples for those who have not carried out a ligand binding assay, or who are trying to develop a ligand binding assay for a novel receptor. Established protocols for many receptors can be found in other references. Protocol 1 describes ligand binding to membranes and Protocol 2 describes ligand binding to whole cells. The advantage of assaying binding to membranes is that this can circumvent a number of experimental difficulties. For example, non-specific effects such as ligand uptake by endocytosis or alteration of the ligand by cellular enzymes can be avoided by using membranes. On the other hand, binding to whole cells offers opportunities to specifically examine the cell surface receptors without interference from internal pools of receptors. This can be important for studying receptor endocytosis or for confirming whether mutant receptors are present at the plasma membrane.

Several general properties of the binding reaction must be established before carrying out detailed ligand binding studies. First of all, the time of incubation must be determined as the receptor properties are typically measured in reactions that have reached equilibrium between free ligand and ligand bound to receptors. This can be determined by measuring the rate of association (Figure 1A), and then choosing a time after which there is no further increase in binding. Note that these studies should be carried out at sub-saturating concentrations of ligand and that the time of incubation should not be too long, as it is critical that the ligand remains stable during the time of incubation and is not converted to an inactive state by cellular enzymes. The time of dissociation should also be determined. Typically, this is done by diluting the binding reaction with an excess of buffer to quickly lower the ligand concentration, and then assaying the amount of ligand that remains bound (Figure 1B). This parameter is important to determine how fast significant amounts of ligand dissociate. Ligands that dissociate quickly must be washed very rapidly to prevent significant loss of binding during the experimental manipulations. Another key property is to ensure that ligand binding is specific by showing that the receptor can be saturated with increasing concentrations of ligand (Figure 1C). In contrast, non-specific binding, often determined by adding an excess of cold ligand, does not show saturation.

Once a binding assay is established, investigators may wish to expand the number of samples that can be analysed to facilitate the screening of different cell types or to analyse a large set of mutants. For these high-throughput screening purposes, ligand binding assays can be automated with the use of machines such as a Brandel Cell Harvestor or with other 96-well tray formats (Whatman, Millipore).

## 2.2 Assaying ligand binding to membrane preparations

This protocol provides general binding conditions appropriate for many ligands and receptors. Specific rates of ligand association and dissociation or the amount of protein in the binding assay have to be determined for each particular receptor.

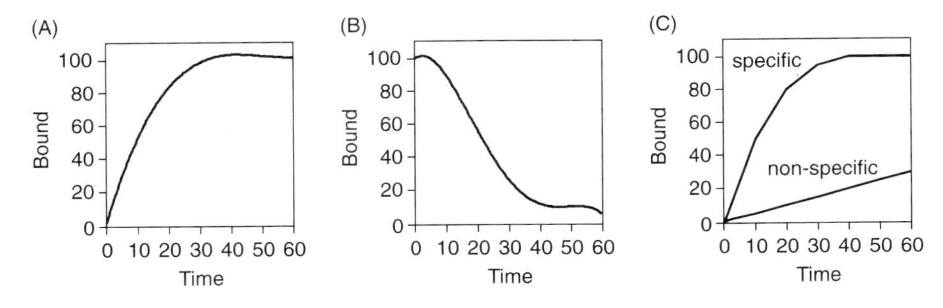


**Figure 1** Conditions for ligand binding assays. (A) In order to determine the time at which ligand binding reaches equilibrium, the rate of ligand association with receptor should be determined. (B) To determine how quickly samples must be washed, the rate of ligand dissociation is measured. (C) The specificity of binding should also be determined. Specific binding of ligand to receptors is expected to reach saturation at high concentrations of ligand. Non-specific binding is not saturated and continues to increase.

## Protocol 1

# Receptor binding assay in membrane preparations

### Equipment and reagents

- Whatman GF/C filters
- Vacuum filtration apparatus (e.g. Fisher vacuum filter holder)
- Cell line expressing the appropriate receptor
- Phosphate-buffered saline (PBS): 137 mM NaCl, 2.7 mM KCl, 8.1 mM $Na_2PO_4$, 1.5 mM $KH_2PO_4$ pH 7.5
- Radiolabelled and unlabelled ligand
- Binding buffer: 50 mM Tris–HCl pH 7.4, 1 mM EGTA, 5 mM $MgCl_2$, 100 μM PMSF, 0.02% bacitracin, 0.001% leupeptin, 0.001% pepstatin A
- 0.1% polyethylenimine (PEI)
- Filter rinse buffer: 50 mM Tris–HCl pH 7.8
- Water-compatible scintillation cocktail (e.g. Fisher ScintiVerse)
- Reagents for protein concentration assay (e.g. Bio-Rad)

### Method

1. Remove growth medium from culture plates containing confluent cells. Rinse cells twice with PBS.
2. Add 10 ml of binding buffer per 500 $cm^2$ plate and scrape cells.
3. Centrifuge cells at 2500 *g* for 10 min at 4 °C. Resuspend in 10 ml of cold binding buffer. Homogenize using Polytron homogenizer.
4. Centrifuge the homogenate at 45 000 *g* for 30 min at 4 °C. Resuspend the membrane pellets in binding buffer.
5. Determine the protein concentration using the protein concentration determination kit, and dilute membranes to 0.5 mg/ml. Store membrane aliquots at –80 °C or proceed to step 6.
6. Prepare two sets of different concentrations of radiolabelled ligand at two times the final concentration in the binding assay. To one add a 100-fold excess of unlabelled ligand.
7. Soak GF/C filters in 0.1% PEI for 1 h at room temperature (or at 4 °C overnight). This reduces non-specific binding and increases the retention of receptor on the glass fibre.
8. Mix 100 μl of membrane preparation (30–50 μg) with 100 μl of ligand. Incubate at room temperature until binding reaches equilibrium. The optimal concentration of membranes in the assay must be determined, as receptor density is variable in different cells.
9. Filter the binding reactions over glass fibre filters pre-soaked in 0.1% PEI. **Caution**: In rapidly equilibrating systems specific binding may be lost during the filter rinses. In these cases, quick washing conditions have to be strictly observed.
10. Rinse three times with 2 ml of cold filter rinse buffer.
11. Determine the total bound and the non-specific bound radioactivity by scintillation counting.

## 2.3 Assaying ligand binding with whole cells

This specific protocol was derived from that described for the binding of α-factor pheromone to its receptor on yeast cells (1), but can be easily adapted to study cell surface receptors on a wide range of cell types. As presented, this protocol describes how to assay the binding of a ligand to one cell type in order to determine binding parameters such as the dissociation constant ($K_d$) and number of total cell surface binding sites ($B_{max}$) at the plasma membrane. Variations of this assay can also be used to screen different cells, such as a set of mutants, using a single concentration of radiolabelled α-factor. Note that *p*-tosyl-L-arginine methyl ester (TAME) is added to the assay to inhibit proteolysis of α-factor and, at the concentration used, does not interfere with its biological activity. Metabolic inhibitors, $NaN_3$ and KF, are added to inhibit energy-dependent processes, such as endocytosis, that affect the number of α-factor binding sites at the plasma membrane. Samples are processed with quick washing conditions because of the rapid dissociation rate of the α-factor.

## Protocol 2

## Binding assay for the yeast α-factor receptor in whole cells

### Equipment and reagents

- Whatman GF/C filters
- Vacuum filtration apparatus (e.g. Fisher vacuum filter holder)
- 1 M $NaN_3$
- 1 M KF
- *p*-Tosyl-L-arginine methyl ester (TAME)
- YPD: 1% yeast extract, 2% bacto-peptone, 2% dextrose
- Filtered IM (inhibitor media) – TAME: YPD, 10 mM $NaN_3$, 10 mM KF
- Filtered IM + 2 × TAME: IM, 20 mM TAME
- Water-compatible scintillation cocktail (e.g. Fisher ScintiVerse)

### Method

1. Set up a 200 ml culture in YPD, and grow cells at 30 °C to log phase ($5 \times 10^6$ to $1 \times 10^7$ cells/ml).
2. Pour the culture into a pre-chilled glass flask and add KF and $NaN_3$ to 10 mM each.
3. Collect cells by centrifugation at 4 °C, 1500 *g*, for 10 min. Wash cells twice by resuspending in 40 ml of ice-cold IM – TAME and pelleting.
4. Resuspend the final cell pellet in IM – TAME to a final volume of 1 ml.
5. Vortex and determine cell concentration under the microscope using a haemocytometer. Dilute cells to $1 \times 10^9$ cells/ml in IM – TAME and keep on ice until needed for the assay.
6. Prepare a set of eight dilutions of [$^{35}$S]α-factor (50 Ci/mmol) in IM + 2 × TAME to give twice the assay concentration. The final assay concentrations should include dilutions that are above and below the $K_d$ (e.g. try from 100 nM to 1 nM), and the ligand should be in at least 10-fold excess over the number of receptors in the assay.

**Protocol 2** continued

7. Split each of the hot α-factor dilutions in two equal portions. Label one set as hot α-factor. Add a 100-fold excess of cold α-factor to the other tube and label this set as cold α-factor.
8. Label four 0.5 ml microcentrifuge tubes for each concentration of α-factor to be assayed (e.g. 1a, 1b, 1c, 1d). Add 50 μl of the cells to the tubes. Label scintillation vials accordingly.
9. Prepare the filter apparatus with one GF/C filter, forceps, a repeat pipetter set for 2 ml, a P200 pipette set for 50 μl, a P200 pipette set for 90 μl, a timer, and IM - TAME on ice. Lay out four GF/C filters for each α-factor concentration so that the filters can be picked up quickly with forceps.
10. Add 50 μl of hot 200 nM α-factor to tube 1a at time zero, and do the same to tube 1b at t = 2 min. Mix well. At 4 min and 6 min add 50 μl of cold 200 nM α-factor to tube 1c and tube 1d respectively. Continue with the next set adding α-factor to cells every 2 min.
11. At time 30 min, take 90 μl of the binding reaction in tube 1a and filter it through a filter apparatus with 2 ml of IM - TAME. Wash twice with 2 ml of IM - TAME, and place filter to dry on top of bench paper. Place another GF/C filter in filter apparatus. Continue to filter reactions sequentially every 2 min.
12. Quantitate radioactivity bound to filters by scintillation counting. Determine specific binding by subtracting the values for the hot from the cold assays. Convert values of radioactivity to amount of ligand bound and plot data.

## 2.4 Data analysis

Because the analysis of binding data can be complex, the use of sophisticated computer programs such as Prism by GraphPad Software is helpful. However, for those new to binding assays, we will describe methods for the initial analysis of binding data that can be carried out using graphing software that is typically available in most laboratories. One of the most common forms of analysis is the Scatchard Plot (Bound Ligand versus Bound/Free; Figure 2A). The Scatchard Plot is popular because it converts binding data into a linear relationship that can be readily used to infer receptor properties. The Bound value is determined from the binding assays as the amount of ligand bound at a given concentration. The amount of Free ligand can be determined experimentally, or by subtracting the Bound from the Total. Note that the assay conditions should be set up so that the Bound ligand corresponds to <10% of the initial ligand concentration to avoid the need for complicated adjustments at concentrations where ligand is limiting. If the results of the Scatchard Plot yield a linear relationship, the number of ligand binding sites can be determined from the $x$ intercept value and the affinity for ligand can be determined from the slope of the line that best fits the data points (slope = $-1/K_d$). To simplify this analysis, the values of B and F were adjusted from the initial assay conditions to that of a 1 litre reaction. Thus, the

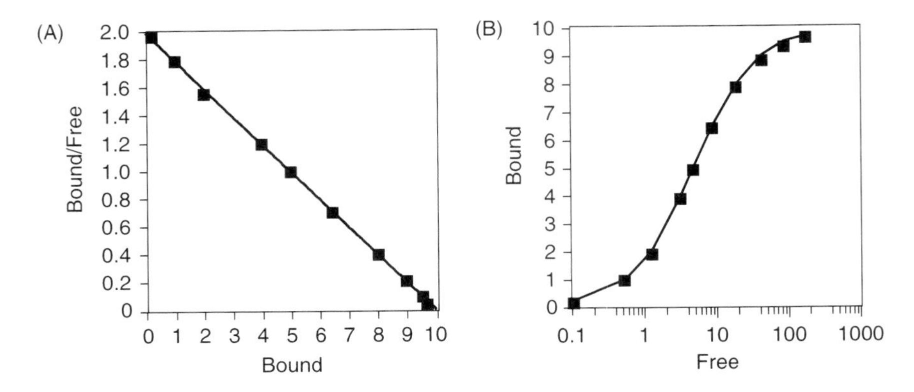


**Figure 2** Graphical analysis of ligand binding data. (A) To determine ligand binding parameters, binding data can be analysed by a Scatchard Plot (Bound/Free versus Bound). As discussed in Section 2.4, the dissociation constant ($K_d$) can be derived from the slope of the line and the number of binding sites can be derived from the *x* intercept. (B) To ensure that a sufficient number of data points have been analysed in the Scatchard Plot, the data should also be plotted as Bound versus the log(Free) to show that an S-shaped curve is obtained.

10 nmol bound at the *x* intercept corresponds to $6 \times 10^{15}$ ligand binding sites present per corresponding number of cells or mg of membrane. The slope of the line is –0.2 corresponding to a $K_d$ of 5 nM. If a linear relationship is not found in the Scatchard Plot, a more detailed reference should be consulted to help determine whether the assay was flawed or if there are complex binding properties such as receptors with different affinities or cooperative effects in binding ligand.

Although Scatchard Plots are a relatively simple way to analyse data, they are often criticized because they can easily be misinterpreted (2). One of the most common mistakes is to analyse an incomplete set of data points. For Scatchard analysis to be meaningful, the data points should span from ligand concentrations that give a low level of occupancy to concentrations which essentially saturate the receptors. To examine this, plot the Bound versus Free as shown in Figure 2B. A complete set of data points is expected to give an S-shaped curve with an inflection occurring at the concentration of ligand that equals the $K_d$. As the development of binding studies progresses, it is advised that the investigator consult other references for information regarding the theoretical background of these graphing methods and their limitations.

# 3 Epitope tags and fusion proteins

## 3.1 Protein detection tags

Epitope tags are used to characterize proteins of interest because they circumvent the need for raising specific antibodies. Some of the sequences commonly employed to epitope tag proteins for immunodetection are listed in Table 1.

**Table 1** Protein detection tags

| Tag | Sequence | Application | Suppliers[a] |
|---|---|---|---|
| Haemagglutinin (HA) | YPYDVPDYA | Immunodetection | BAbCO, Roche, Zymed, Clontech, Santa Cruz |
| c-Myc | EQKLISEEDL | Immunodetection | Santa Cruz, Stratagene, Clontech, Invitrogen, Roche |
| FLAG | DYKDDDDKI | Immunodetection | Stratagene, Santa Cruz, Sigma |
| T7 | MASMTGGQQMG | Immunodetection | Novagen, Clontech, Invitrogen, Santa Cruz, Roche, Chemicon, Zymed |
| GFP | 236 amino acids[b] | Visualization | Clontech, Invitrogen, Santa Cruz, Roche, Chemicon, Zymed |
| 6 × His | HHHHHH | Purification | Invitrogen, Clontech, Santa Cruz, Roche |

[a] Partial list of companies that supply antibodies and/or expression vectors for a particular tag.

[b] GFP is the *Aequoria victoria* green fluorescent protein. Blue, yellow, cyan, and enhanced fluorescence variants have been developed and are commercially available.

Included are the haemagglutinin (HA), c-myc, FLAG, and T7 tags. The HA, myc, and FLAG have been used extensively to tag receptor proteins both at the N terminus and C terminus. Other commonly used tags are the hexahistidine (6 × His) and green GFP. Hexahistidine-fusion proteins facilitate rapid protein purification by chromatography on Ni-affinity columns. GFP fusions at the N terminus or C terminus of receptors are employed to visualize the proteins in living cells and for FACS analysis (3). GFP and the blue (BFP), cyan (CFP), or yellow (YFP) variants can be used in combination for tagging two proteins and study co-localization by confocal microscopy or protein–protein interactions by fluoresce resonance energy transfer (FRET).

## 3.2 Chimeric receptors and GPCR-Gα fusion proteins

A widely used approach to study receptor function has been the analysis of hybrid receptors constructed between functionally distinct members of a receptor family. Tyrosine kinase receptor chimeras have been very useful in defining functionally separable receptor regions that determine ligand binding and activation of intracellular signals. In the case of GPCRs, the ligand binding and receptor signalling domains are not as clearly separable as in tyrosine kinase receptors. However, chimeric receptor approaches are useful in identifying structural determinants for ligand specificity and receptor–G protein coupling selectivity (4). In addition, GPCR-Gα fusion proteins have been used to minimize the influence of receptor–G protein density or restricted localization of G proteins at the plasma membrane (5). This type of fusion ensures close physical proximity, establishes a defined 1:1 stoichiometry, and promotes pre-coupling of the GPCR and the Gα subunit. These properties make GPCR-Gα chimeras useful tools for the molecular analysis of receptor–G protein coupling and for improving GPCR function in heterologous systems.

# 4 Receptor modifications and receptor structure

## 4.1 Post-translational modifications

Post-translational modifications play important roles in regulating the function and stability of receptor proteins. Some of the commonly observed modifications include glycosylation, phosphorylation, and ubiquitination. Protocols will be described in this section for the analysis of post-translational modifications using convenient gel mobility shift assays. For example, glycosylation and phosphorylation can often be monitored by treating samples with endoglycosidases or phosphatases, and then the presence or absence of modification can be inferred by the mobility of the receptor proteins as visualized by Western blotting (Protocol 3). Other modifications can be studied with specific antibodies. For example, ubiquitination can be directly monitored with anti-ubiquitin antibodies (Novocastra Laboratories Ltd., Zymed Laboratories Inc.) or by co-expressing HA- or myc-tagged ubiquitin and the receptor of interest. In addition, phosphorylation on tyrosine residues can also be studied with highly specific and sensitive anti-phosphotyrosine antibodies that are commercially available (Upstate Biotechnology Inc., Santa Cruz Biotechnology Inc.). Unfortunately, antibodies raised against phosphoserine and phosphothreonine do not seem to work nearly as well.

The following protocol for the analysis of post-translational modifications of the yeast α-factor receptor can be adapted to analyse a wide range of receptors in different cell types. In addition, investigators studying receptors in animal cells can extend these methods by comparing the effects of different endoglycosidases. Whereas all *N*-linked glycosylation in yeast is expected to be sensitive to Endo H, only immature forms of receptors are expected to be Endo H sensitive in animal cells. The reason for this is that the initial high mannose groups that are added to proteins are further modified in animal cells so that they are no longer cleaved by Endo H, but can be cleaved by other enzymes such as PNGase.

## Protocol 3

## Endoglycosidase H (Endo H) and λ protein phosphatase treatment of the yeast α-factor receptor

### Equipment and reagents

- Acid treated glass beads (Sigma Chemical Company, Cat. No. G8772)
- Endoglycosidase H (New England Biolabs)
- λ protein phosphatase (New England Biolabs)
- Endo H buffer: 100 mM NaCl, 50 mM Tris–HCl pH 7.5, 1 mM EDTA
- λPP buffer: 50 mM Tris–HCl pH 7.5, 0.1 mM EDTA, 5 mM DTT, 0.1% Frij35
- Cycloheximide
- 1 M $NaN_3$ and 1 M KF
- TE/PP: 50 mM Tris–HCl pH 7.5, 1 mM EDTA, 100 μg/ml PMSF, 2 μg/ml pepstatin A
- Urea sample buffer: 1 g of urea in 1 ml of 17.5 mM Tris–HCl pH 6.8, 1.75% SDS, 1% 2-ME, 0.1% bromophenol blue
- Reagents to determine protein concentration (e.g. Bio-Rad)

**Protocol 3** continued

### Method

1. Pre-chill solutions, tubes, and centrifuges to 4 °C.
2. Harvest a 50 ml culture of exponentially growing cells.
3. Add cycloheximide (20 μg/ml) and incubate at 30 °C for 15 min. (This is an optional step to block new receptor synthesis).
4. Add α-factor ($1 \times 10^{-7}$ M) and incubate again at 30 °C for 15 min.
5. Collect cells by centrifugation at 1500 *g* in pre-chilled tubes containing 0.5 ml of 1 M $NaN_3$ and 0.5 ml of 1 M KF.
6. Wash cells twice with 10 ml TE buffer and once with 1 ml TE/PP buffer.
7. Resuspend in 0.6 ml TE/PP.
8. Add glass beads and lyse the cells by vigorous vortexing four times for 1 min each time.
9. Spin at 330 *g* for 5 min at 4 °C.
10. Collect 0.5 ml of supernatant and transfer it to a fresh microcentrifuge tube.
11. Centrifuge at 12 000 *g* for 20 min in a microcentrifuge, and resuspend the membrane pellet in Endo H buffer or λPP buffer depending on the treatment.
12. Determine protein concentration and dilute to a final concentration of 5 mg/ml.
13. Incubate membranes with either Endo H or λ protein phosphatase as follows:
    (a) To study glycosylation, incubate 10 μl (50 μg) of membranes in Endo H buffer with 3 μl (3000 U) of Endo H at 37 °C for 2 h.
    (b) To study phosphorylation, incubate 10 μl (50 μg) of protein in phosphatase buffer containing 2 mM $MnCl_2$ and 100–200 U of λ protein phosphatase at 30 °C for 1 h.
14. Add 30 μl of urea sample buffer and analyse by Western blot.

## 4.2 Receptor structure

Many different methods can be used to probe receptor structure, so we will limit our comments here to a few techniques that, like the protocol described above, take advantage of convenient gel mobility shift assays. One method that works well in this format is to examine receptor structure after protease digestion. The resulting proteolytic fragments can be resolved on gels to map the domain containing a particular post-translational modification or to map membrane topology. This form of analysis can be greatly enhanced by using mutagenesis strategies to introduce sites for specific proteases such as Factor X (Ile–Glu/Asp–Gly–Arg) and the TEV (Glu–Asn–Leu–Tyr–Phe–Gln–Gly) that have unusual recognition sites that are rarely encountered in native proteins.

Gel mobility shift assays have also been used in combination with chemical crosslinking to define intra- and intermolecular contacts in receptor proteins. In some strategies, exogenous chemical crosslinkers are used to covalently bond proteins together. To achieve a finer degree of resolution, cysteine residues can be engineered into the receptor protein and assayed for the ability to form disulfide bonds after oxidation (6). This method is gaining increased popularity for mapping contacts between the transmembrane segments of G protein-coupled receptors (7). Intermolecular crosslinking is easily detected as this results in a dimer that migrates with slower gel mobility than the monomeric proteins. In contrast, intramolecular crosslinking is more difficult to detect as it does not always result in altered gel mobility. To circumvent this problem, split receptors are generated either by proteolysis or by co-expressing two halves of a receptor in the same cell. The split receptors are then assayed by Western blotting to determine whether the halves of the receptor run in their monomer position, or if they run with a slower mobility indicative of an intramolecular crosslink.

# 5 Receptor trafficking

## 5.1 Visualization of receptor trafficking and analysis of receptor loss from the plasma membrane

Double-label immunocytochemistry has been widely used to monitor receptor trafficking within the cell. However, the use of receptor–GFP fusions has gained popularity in recent years. A wide variety of receptors fused to GFP exhibit normal agonist binding affinity, receptor number at the plasma membrane, and internalization rates (3). These properties make GFP fusions a popular technique that allows real time visualization of receptor trafficking to and from the plasma membrane. As an example, Figure 3 shows the localization of GFP-tagged yeast α-factor receptor in the absence of α-factor and upon α-factor stimulation. Note that prior to stimulation, the receptor–GFP fluorescence is found diffusely distributed over the surface of the cell. In contrast, the cells treated with α-factor show that the receptor–GFP has undergone a redistribution and is now found clustered into endosomal compartments.

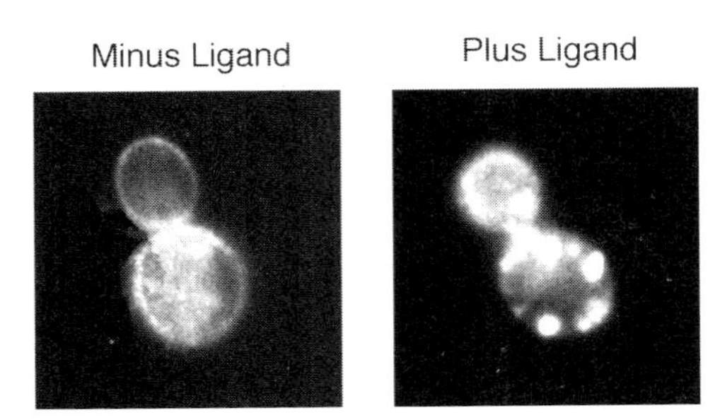


**Figure 3** Visualization of GFP-tagged α-factor receptors. Fluorescent microscope images of yeast carrying a plasmid that encodes an α-factor receptor with GFP-tagged at the C terminus. Exponentially growing cells were incubated in the absence or presence of $5 \times 10^{-7}$ M α-factor for 20 min and then photographed.

Other methods can be used to measure indirectly the rate of receptor internalization. For example, the down-regulation of receptor binding sites from the plasma membrane can be determined quantitatively by following the loss of receptors at times after ligand addition. For these experiments it may be necessary to block new receptor synthesis with cycloheximide. Cells are then collected in the presence of metabolic inhibitors to prevent further endocytosis. The bound ligand is removed by washing and then the receptor number at the plasma membrane is determined by a ligand binding assay (Protocol 2). Alternatively, ligand-induced receptor endocytosis can be measured by incubating cells with radiolabelled ligand, washing to remove extracellular ligand, and determining the cell-associated radioactivity.

## 5.2 Membrane fractionation

Membrane fractionation is also commonly used to monitor internalization and trafficking of receptors. Here we provide two protocols for membrane fractionation. The first is a general protocol for fractionation of membranes on sucrose gradients (Protocol 4). The second protocol has been adapted from Schandel and Jenness (8), and describes fractionation of yeast cell membranes on Renocal gradients (Protocol 5). Both methods allow the separation of plasma membranes from more buoyant internal membranes, such as endoplasmic reticulum, Golgi complex, and vacuole. However, Renocal gradients optimize the separation of heavier plasma membrane fractions from other membranes. Renocal gradient fractionation is carried out at high ionic strength so this technique is suitable for the analysis of integral membrane proteins and proteins tightly associated to the plasma membrane, but it is not recommended for proteins peripherally associated with the plasma membrane.

## Protocol 4

### Membrane fractionation by sucrose density gradient centrifugation

#### Equipment and reagents

- Ultracentrifuge tubes for Beckman SW41 rotor
- PBS: 137 mM NaCl, 2.7 mM KCl, 8.1 mM $Na_2PO_4$, 1.5 mM $KH_2PO_4$ pH 7.5
- Lysis buffer: 10 mM Tris–HCl pH 7.4, 1 mM EDTA
- Sucrose solutions: 55–30% sucrose in 2 mM Tris–HCl pH 8.0
- 4 × SDS–PAGE loading buffer: 200 mM Tris–HCl pH 6.8, 400 mM DTT, 8% SDS, 0.4% bromophenol blue, 40% glycerol

#### Method

1. Pre-chill reagents, tubes, and centrifuges.
2. Grow cells in 150 mm diameter dishes to semi-confluency.

**Protocol 4** continued

3 Wash cells three times with PBS. Add 1 ml of hypotonic lysis buffer to each plate and harvest cells with the help of a rubber policeman.

4 Combine samples of cell lysate, vortex for 1 min, and keep on ice while preparing the discontinuous sucrose gradients.

5 Gently place 0.9 ml of each sucrose solution in an ultracentrifuge tube. From bottom to top add 55%, 50%, 47.5%, 45%, 42.5%, 40%, 37.5%, 35%, 32.5%, and 30% sucrose.

6 Overlay each tube with 1 ml of cell lysate.

7 Spin at 110 000 g (30 000 rpm) in Beckman SW41 rotor for 3 h at 4 °C.

8 Label two sets of microcentrifuge tubes. To one set add 25 μl of 4 × SDS–PAGE loading buffer.

9 Collect twenty 0.5 ml fractions. Use a pipette to remove successive fractions from the top of the gradient to the bottom. Place samples in a microcentrifuge tube.

10 Transfer 75 μl of each fraction to tubes containing 4 × SDS–PAGE loading buffer. Keep the rest of each fraction at −80 °C until needed.

11 Run aliquots on a polyacrylamide gel and analyse by Western blot. Plasma membranes should be contained in fractions 10–15. Detect plasma membrane and internal membranes by probing blots with antibodies against specific marker proteins.

## Protocol 5

## Membrane fractionation by Renocal density gradient centrifugation

### Equipment and reagents

- Ultracentrifuge tubes for Beckman SW50.1 rotor
- Renocal-76, 37% solution (Bracco Diagnostics)
- 50 mM PMSF in ethanol
- Pepstatin A: 1 mg/ml in ethanol
- TE buffer: 50 mM Tris–HCl pH 7.5, 1 mM EDTA
- TE/PP buffer: TE, 100 μg/ml PMSF, 2 μg/ml pepstatin A
- Urea sample buffer (see Protocol 3)
- Renocal solutions: 34–22% Renocal-76 in TE

### Method

1 Pre-chill solutions, tubes, and centrifuges.

2 Harvest 50 ml culture of log phase yeast cells at a density of $<10^7$ cells/ml. Collect cells by centrifugation at 1500 g. If treated with α-factor, add $NaN_3$/KF to 10 mM final concentration.

**Protocol 5** continued

3. Wash cells twice with 10 ml TE buffer and once with 1 ml TE/PP buffer. Resuspend in 0.65 ml TE/PP.
4. Add glass beads and lyse the cells by vigorous vortexing four times for 1 min each time. Chill on ice between intervals of vortexing.
5. Spin at 330 g for 5 min at 4 °C.
6. Combine 0.55 ml of supernatant with 0.55 ml of 37% Renocal-76 solution. Place 1 ml of diluted lysate at the bottom of SW50.1 ultracentrifuge tube.
7. Form a Renocal step gradient by sequentially adding the following solutions very gently: 1 ml of 34% Renocal, 1 ml of 30% Renocal, 1 ml of 26% Renocal, 1 ml of 22% Renocal
8. Spin in Beckman SW50.1 rotor at 150 000 g for 20 h, at 4 °C.
9. Carefully collect 360 μl fractions from the top of the tubes in microcentrifuge tubes containing 5 μl of 50 mM PMSF and 0.7 μl of 1 mg/ml pepstatin A.
10. Dilute part of each fraction 1:3 with urea sample buffer. Freeze the rest at −70 °C.
11. Heat samples at 37 °C for 10 min and load 20 μl on a 10% acrylamide gel. Analyse by Western blot. Internal membranes are separated in fractions 2–6, plasma membrane in fractions 9–11, and cytosolic proteins in fractions 13 and 14 (Figure 4).
12. Probe the blots with antibodies for proteins expressed in specific membrane compartments. Among others, plasma membrane ATPase (Pma1), HDEL, and G6PDH antibodies are appropriate as markers for plasma membrane, internal membranes and cytosol, and cytosol respectively (8, 9).

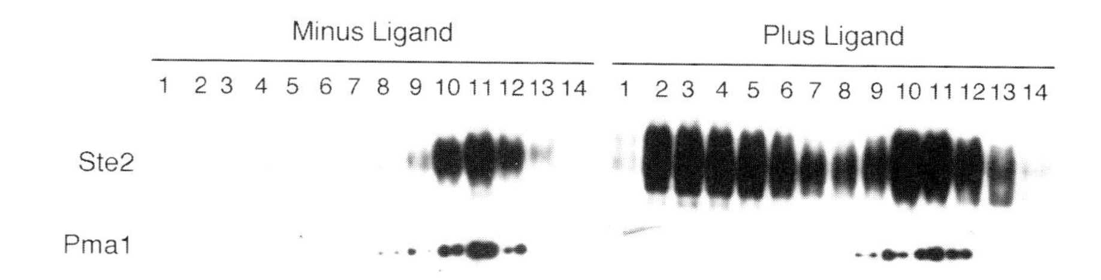


**Figure 4** Subcellular localization of yeast α-factor receptors by Renocal density centrifugation. Exponentially growing yeast cells were incubated in the absence (left panel) or presence (right panel) of $5 \times 10^{-7}$ M α-factor for 2 h. Membranes were resolved on Renocal-76 density gradients as described in Protocol 5. An equivalent amount of each gradient fraction was analysed on a Western blot probed with anti-Ste2 antibodies to detect the α-factor receptors. For comparison, the fractions were also probed with an anti-Pma1 monoclonal antibody that detects the plasma membrane ATPase to show the fractionation of plasma membrane proteins.

# 6 Heterologous expression of receptors in yeast

## 6.1 Background

Yeast cells are a useful expression system for receptors from other cell types because its experimental accessibility can be exploited for a wide array of genetic, molecular biology, and high-throughput screening assays. Another advantage that is particularly relevant for studies on G protein-coupled receptors is that it has been possible to achieve functional expression of mammalian receptors in yeast (10, 11). The basic outline of these studies is that a foreign receptor is expressed in place of the G protein-coupled mating pheromone receptors that promote conjugation in yeast. The capacity of heterologous receptors to function in yeast is then assessed by analysing their ability to activate the mating pheromone signal pathway. This strategy has been beneficial for the functional analysis of receptors in the absence of other receptor subtypes or for other applications such as screening for agonists and antagonists.

## 6.2 General considerations

The first step in expressing a foreign receptor in yeast is to choose an expression vector. Yeast expression vectors are 'shuttle' plasmids which can be propagated and stably maintained in both yeast and bacteria. To select for maintenance in yeast, these plasmids carry a marker gene that complements the auxotrophy of yeast strains with mutations in that gene. The three main classes of plasmids are YIp (yeast integrating plasmid), YCp (yeast centromeric plasmid), and YEp (yeast episomal plasmid). YIp plasmids are stably integrated in the genome as a single copy per cell, YCp plasmids replicate autonomously at a low copy number, and YEp plasmids replicate at very high copy numbers. Commonly used strong promoters for the constitutive expression of genes in yeast are derived from the genes encoding alcohol dehydrogenase (*ADH*) or glyceraldehyde-3-phosphate dehydrogenase (*GPD*). The promoter from the galactokinase (*GAL1*) is commonly used as an inducible promoter. Vectors carrying these promoters with a variety of selectable markers have been described (12, 13) and are available from the American Type Tissue Culture Collection (ATCC).

Another important consideration is the choice of a yeast strain that has been modified to increase the sensitivity for detection of signalling by a foreign receptor (10). A haploid strain of either of the two different mating types (**a** or α) must be used as the **a**/α diploid cells do not respond to pheromone (for a review of the pheromone signal pathway see ref. 14). If a *MAT**a*** strain is employed, it is crucial that the α-factor receptor gene (*STE2*) is deleted to prevent interference. Similarly, the **a**-factor receptor gene (*STE3*) should be deleted if a *MATα* strain is used. Although some mammalian receptors are able to interact with the endogenous G protein, many receptors will only couple to chimeric G proteins in which the C terminal sequences of the yeast Gα protein (*GPA1*) are substituted with sequences from the corresponding mammalian Gα (10). In addition, to further improve the efficiency of signalling, the gene for the RGS protein (*SST2*) that down-regulates the G protein signal can be deleted to prolong activation of the pathway. The *FAR1* gene should also be deleted to avoid cell division arrest in response to receptor activation. Finally, a pheromone-responsive reporter gene such as *FUS1-lacZ* or *FUS1-HIS3* should be introduced to detect receptor signalling.

## 6.3 Growth and transformation of yeast

Yeast cells can be grown in rich non-selective (YPD) medium or synthetic dextrose minimal (SD) medium (15). These media can be prepared in the laboratory (15) or purchased from Bio 101. The most common method for the introduction of DNA into yeast uses lithium acetate (16) and is described in Protocol 6.

### Protocol 6

### Lithium acetate transformation

#### Reagents

- Salmon sperm carrier DNA: salmon testes DNA (Sigma) dissolved at 20 mg/ml in 10 mM Tris–HCl pH 8.0, 1 mM EDTA, sonicated, and boiled
- Lithium acetate/TE: 10 mM Tris–HCl pH 7.5, 1 mM EDTA, 100 mM lithium acetate
- PEG/lithium acetate/TE: lithium acetate/TE, 40% polyethylene glycol 4000
- YPD medium: 1% yeast extract, 2% bacto-peptone, 2% dextrose, 0.006% adenine
- SD medium: 0.67% bacto-yeast nitrogen base, 2% glucose, and amino acids with the exception of the one used for selection; add 1.8% bacto-agar for solid plates

#### Method

1. Inoculate 5 ml of YPD with one colony of the yeast strain to be transformed, and grow overnight at 30 °C.
2. In the morning, inoculate 50 ml of YPD with the overnight culture to a density of $2.5 \times 10^6$ cells/ml. Grow at 30 °C until cells reach 5–10 $\times 10^6$ cells/ml (approx. 4 h).
3. Harvest cells by centrifugation at 1500 *g* for 5 min. Wash the cell pellet in 25 ml of sterile water, and then 5 ml of lithium acetate/TE. Centrifuge cells between washes.
4. Resuspend the pellet in 0.5 ml of lithium acetate/TE. Cells can be used immediately for transformation or stored for up to two days at 4 °C.
5. Add 2 μl of carrier DNA, 100–500 ng of plasmid DNA, and 50 μl of cells to a 1.5 ml microcentrifuge tube.
6. Add 0.4 ml of PEG/lithium acetate/TE and vortex.
7. Incubate at 30 °C for 30 min. Heat shock for 15 min at 42 °C.
8. Add 1 ml of sterile water, mix, and spin in a microcentrifuge for 10 sec.
9. Resuspend pellet in 0.2 ml of selective medium.
10. Plate 0.1 ml of transformed cells onto appropriate SD plates. (Save the remainder of the cells at 4 °C as a back-up in case of contamination with mould, etc.)
11. Incubate plates at 30 °C for two days to allow colonies to form.

## 6.4 Analysis of receptor expression and activation

Receptor expression can be analysed by Western immunoblot, using either specific antibodies or antibodies that recognize an epitope tag (see methods described in Protocols 4 and 5). Receptor activation is monitored using reporter genes carrying the pheromone-inducible *FUS1* promoter (17). Activation of the *FUS1-HIS3* reporter gene is measured by assaying the ability of cells to grow in medium lacking histidine and containing different concentrations of 3-aminotriazol, an inhibitor of *HIS3* function. Thus, dose–response assays can be carried out by determining the ability of cells stimulated with increasing concentrations of ligand to grow in media containing different levels of 3-aminotriazol. Alternatively, the *FUS1-lacZ* reporter gene, which leads to receptor-induced production of β-galactosidase, can be used to quantitate receptor signalling by a quantitative colorimetric assay for β-galactosidase activity (18) for use with permeabilized yeast cells. A method for measuring reporter gene activity is described in Protocol 7.

### Protocol 7

### *FUS1-lacZ* (β-galactosidase) reporter gene assay

#### Reagents

- Z buffer: 10 mM KCl, 1 mM $MgSO_4$, 60 mM $Na_2HPO_4$, 40 mM $NaH_2P_4$, 50 mM 2-mercaptoethanol (2-mercaptoethanol is added just prior to use)
- $CHCl_3$
- ONPG (*o*-nitrophenyl-β-D-galactopyranoside): 4 mg/ml in 0.1 M phosphate buffer pH 7.0
- 0.1% SDS
- 1 M $Na_2CO_3$

#### Method

1. Inoculate yeast culture and grow overnight at 30 °C. Adjust culture conditions so that the cells remain in logarithmic phase (less than $10^7$ cells/ml). A good way to do it is to grow an overnight culture in 5 ml to saturation, and the following day inoculate a second overnight with this saturated culture (1:20 to 1:500 dilution) to have the cells exponentially growing the next morning.
2. Adjust the culture to $2 \times 10^6$ cells/ml and set a 2 ml culture to grow at 30 °C for 2 h.
3. To measure activation of the α-factor receptor (*STE2*) add α-factor to $1 \times 10^{-7}$ M and incubate for 2 h. This step may be substituted by incubation with the ligand for the receptor analysed, or skipped if only basal signalling activity is going to be measured.
4. Determine $OD_{600}$ for the cell culture.
5. Transfer 0.1 ml of the culture to a 1.5 ml tube containing 0.7 ml Z buffer (with freshly added 2-mercaptoethanol), 50 μl chloroform, 50 μl of 0.1% SDS, and 160 μl of ONPG. Vortex vigorously for 1 min.

**Protocol 7** continued

6 Incubate at 37 °C for 1 h.

7 Quench the reactions with 0.4 ml of 1 M $Na_2CO_3$.

8 Spin the tubes for 10 min in a microcentrifuge to remove debris, and then read absorbance at $OD_{420}$.

9 Calculate β-galactosidase units using the formula:

$$1000 \times OD_{420}/t \times V \times OD_{600}$$

where t = the time of incubation in min, and V = the volume of cell culture added to Z buffer in ml.

## References


1. Jenness, D. D., Burkholder, A. C., and Hartwell, L. H. (1983). *Cell*, **35**, 521.
2. Klotz, I. M. (1982). *Science*, **217**, 1247.
3. Tsien, R. Y. (1998). *Annu. Rev. Biochem.*, **67**, 509.
4. Kobilka, B. K., Kobilka, T., Daniel, K., Regan, J. W., Caron, M. G., and Lefkowitz, R. J. (1988). *Science*, **240**, 1310.
5. Seifert, R., Wenzel-Seifert, K., and Kobilka, B. K. (1999). *Trends Pharmacol. Sci.*, **20**, 383.
6. Chervitz, S. A. and Falke, J. J. (1995). *J. Biol. Chem.*, **270**, 24043.
7. Yu, H., Kono, M., McKee, T., and Oprian, D. (1995). *Biochemistry*, **34**, 14963.
8. Schandel, K. A. and Jenness, D. D. (1994). *Mol. Cell. Biol.*, **14**, 7245.
9. Dosil, M., Giot, L., Davis, C., and Konopka, J. B. (1998). *Mol. Cell. Biol.*, **18**, 5981.
10. Pausch, M. H. (1997). *Trends Biotechnol.*, **15**, 487.
11. Broach, J. R. and Thorner, J. (1996). *Nature*, **384**, 14.
12. Mumberg, D., Muller, R., and Funk, M. (1994). *Nucleic Acids Res.*, **22**, 5767.
13. Mumberg, D., Muller, R., and Funk, M. (1995). *Gene*, **156**, 119.
14. Leberer, E., Thomas, D. Y., and Whiteway, M. (1997). *Curr. Opin. Genet. Dev.*, **7**, 59.
15. Sherman, F. (1991). In *Methods in enzymology* (eds. C. Guthrie and G. Fink), Vol. 194, pp. 3–21. Academic Press, London.
16. Schiestl, R. H. and Gietz, R. D. (1989). *Curr. Genet.*, **16**, 339.
17. Trueheart, J., Boeke, J. D., and Fink, G. R. (1987). *Mol. Cell. Biol.*, **7**, 2316.
18. Miller, J. H. (1972). In *Experiments in molecular genetics*, pp. 325–55. Cold Spring Harbor Laboratory Press, Cold Spring Harbor, New York.

# Chapter 10
# Measurement of signal transduction machinery

Matthew Hodgkin
Department of Biological Sciences, University of Warwick, Coventry CV4 7AL, UK.

Michael Wakelam and Carolyn Armour
The Cancer Research Campaign Institute for Cancer Studies, The University of Birmingham, Clinical Research Block, Edgbaston, Birmingham B15 2TA, UK.

## 1 Introduction

We describe some of the common methods used to assay signal transduction events in cells. The methods presented quantify both signal transduction enzyme activity and the production of second messenger molecules. In view of the wide variety and complexity of signal transduction pathways that are currently understood, some pathways and enzymes may not be covered in great detail or indeed at all. Therefore, where possible, we have attempted to illustrate how the presented protocols can be adapted to other systems.

## 2 Quantification of phospholipase C (PLC) signalling in cells

In mammalian cells, receptor occupation stimulates PLC-catalysed hydrolysis of a minor phospholipid, phosphatidylinositol 4,5-bisphosphate (PtdIns(4,5)$P_2$) to yield Ins(1,4,5) trisphosphate (Ins(1,4,5)$P_3$) and diacylglycerol (DAG). Ins(1,4,5)$P_3$ releases calcium from intracellular stores, which together with the elevated DAG causes activation of protein kinase C (PKC) and physiological responses. Ins(1,4,5)$P_3$ is rapidly dephosphorylated in stages to inositol by specific phosphatases (some of which are inhibited by lithium). Inositol is recombined with DAG (in the form of CMP-PA) to complete the PI cycle (1). Thus quantification of inositol phosphate production, calcium release, and PKC activation can represent quantitative measures of PLC-regulated processes.

## 2.1 Measurement of PLC signal transduction pathway activity

The relatively small amount of inositol-containing lipids in cells means that measurement of PLC activity requires isotopic labelling of the precursor polyphosphoinositides. The simplest method is to label the endogenous pool of polyphosphoinositides with *myo*-[$^3$H]inositol and then quantify the production of *myo*-[$^3$H]inositol phosphates. The use of *myo*-[$^3$H]inositol to label polyphosphoinositides to measure PLC activity offers several advantages over other lipid labelling methods that utilize, for example, [$^{32}$P]Pi, labelled fatty acids, or labelled glycerol. *Myo*-inositol is specifically incorporated into the lipids and the tritium is not lost from the inositol, thus the soluble inositol phosphates produced are also labelled. By definition, fatty acids and glycerols cannot label the soluble inositol phosphates and will label most lipids including polyphosphoinositides. [$^{32}$P]Pi labels all lipids, together with nucleotide phosphates and sugar phosphates, and although this isotope can be extremely useful, mCi quantities are required. However, to label the inositol pool in cells to steady-state or isotopic equilibrium may take at least 48 hours and in practice in the laboratory it is considerably safer to do this with tritium. It should be noted that it is only after a prolonged labelling period that any changes in the amount of radioactivity in a given [$^3$H]inositol-labelled molecule can be assumed to be equivalent to a change in mass. This, however, does not discount the use of shorter labelling periods to indicate the presence of a PLC activity.

The method described in Protocol 1 relies on the presence of lithium before and during the stimulation of the cells to trap inositol phosphates before they are fully dephosphorylated to inositol (see above). However, the lithium trap cannot prevent the first phosphatase in the degradation pathway (the Ins(1,4,5)$P_3$ 5-phosphomonoesterase) from dephosphorylating Ins(1,4,5)$P_3$ to Ins(1,4)$P_2$. Thus, in order to observe differences in the amount of labelled Ins(1,4,5)$P_3$ between basal and stimulated conditions, several time points within the first minute of stimulation may be required.

## Protocol 1

## Pre-labelling cells in culture with *myo*-[$^3$H]inositol

### Reagents

- *Myo*-[$^3$H]inositol in aqueous solution: 20 mCi/mmol, 1 mCi/ml (Amersham)
- Labelling medium: normal growth medium without inositol and serum (e.g. inositol-free DMEM) but containing 10 $\mu$Ci/ml *myo*-[$^3$H]inositol and 0.1–1% BSA
- Post-labelling medium: a suitable cell growth medium (e.g. DMEM) containing 10 mM inositol and 10 mM LiCl
- Stopping solution: 15% (v/v) perchloric acid
- Neutralizing solution: 0.1 M MES, 2 M KOH, 100 mM EDTA
- Universal indicator solution

**Protocol 1** continued

### Method

1. Grow cells to confluence in normal growth medium (12- or 24-well plates will give sufficient material for analysis).
2. Remove growth medium and replace with labelling medium for 48 h.

   **Note**: if the cells cannot be maintained in BSA and require serum, the endogenous inositol in serum should be removed by dialysis and the serum sterile-filtered. For 12-well plates, between 0.5–1 ml of labelling medium should be used.
3. After 48 h, remove the labelling medium and wash the cells twice with minimal medium to remove excess inositol.
4. Incubate the cells for 15 min with post-labelling medium (0.5–1 ml). The unlabelled inositol begins to chase out the [$^3$H]inositol and the lithium chloride sets the trap.
5. After the 15 min pre-incubation, stimulate the cells in the lithium-containing medium. (Prepare a 100 × stock solution agonist, dilute to 10 × in medium, and add 55–500 μl in the well to give a final agonist concentration of 1 ×.)
6. To quench the reaction, add ice-cold 15% perchloric acid to a final concentration of 5%, and transfer the tissue culture plate to ice. Adherent cells should now be removed using a scraper, the debris quantitatively transferred to a microcentrifuge tube, and incubated on ice for 30 min. Non-adherent labelled cells can be transferred to microcentrifuge tubes prior to stimulation and then acid-quenched on ice directly. Microcentrifuge the acid extracts at full speed for 5 min to precipitate protein and membrane material.
7. Neutralize the acid extracts by adding 0.1 M MES, 2 M KOH, 100 mM EDTA. The exact amount of KOH required should be titrated beforehand using universal indicator solution. Universal indicator should not be included when neutralizing the samples as it can interfere with subsequent analysis. The samples should be incubated on ice for a further 30 min, and centrifuged in a microcentrifuge to precipitate perchlorate salts. The neutralized supernatants can be transferred to clean tubes and stored at −20 °C prior to analysis.

### 2.1.1 Analysis of acid extracted inositol phosphates

The labelled inositol phosphates in the extracts can be quantified in one of two ways. A routine low resolution method utilizes Dowex anion exchange resin AG1X-8 (200–400 mesh) in the formate form (available from Bio-Rad). The low resolution technique described in Protocol 2 is high-throughput, requires little special equipment, and is useful as an initial screen, but cannot absolutely identify changes in Ins(1,4,5)$P_3$ mass because it can only resolve inositol phosphates into classes (i.e. InsP, Ins$P_2$s, Ins$P_3$s, Ins$P_4$s, and Ins$P_5$s with Ins$P_6$).

## Protocol 2

## A simple protocol for separation of inositol phosphates

### Equipment and reagents

- Scintillation counter
- Dowex AG1X-8 resin (200–400 mesh in the formate form)
- Scintillation cocktail
- 60 mM ammonium formate/5 mM sodium tetraborate
- 0.2–2 M ammonium formate/0.1 M formic acid

### Method

1. Wash the Dowex AG1X-8 resin in large amounts of deionized water.
2. Aliquot the washed resin into 15 ml tubes to give a bed volume of 0.75 ml. An alternative approach is to aliquot the resin into 10 ml disposable plastic protein purification columns (Bio-Rad).
3. Dilute the extracts with 5–10 ml of water and apply to the resin. For the batch method, mix the beads thoroughly with the sample for 5 min and allow to settle. Centrifuge the beads at low speed (1000 g, 5 min) and retain the supernatant for scintillation counting. For the column method, allow the diluted sample to drip through the column under gravity. Retain the 10 ml of water for scintillation counting. Under these conditions, the inositol phosphates bind to the Dowex resin whereas free labelled inositol is in the water fraction.
4. Elute the bound phosphates using the following stepwise gradient and for both the batch or column methods repeat the procedure as described in step 3. Keep the eluates and scintillation count.
   (a) 10 ml of 60 mM ammonium formate/5 mM sodium tetraborate elutes glycerophosphoinositol.
   (b) 10 ml of 0.2 M ammonium formate/0.1 M formic acid elutes inositol monophosphates.
   (c) 10 ml of 0.4 M ammonium formate/0.1 M formic acid elutes inositol bisphosphates.
   (d) 10 ml of 0.8 M ammonium formate/0.1 M formic acid elutes inositol trisphosphates.
   (e) 10 ml of 1.2 M ammonium formate/0.1 M formic acid elutes inositol tetrakisphosphates.
   (f) 10 ml of 2 M ammonium formate/0.1 M formic acid elutes inositol pentakis- and hexakisphosphates.

### 2.1.2 High resolution separation of inositol phosphates by HPLC

Alternatively, a high resolution anion exchange HPLC method, such as that described in Protocol 3, can be used to separate inositol phosphates. In particular, this method will separate the common $InsP_2$, $InsP_3$, and $InsP_4$ isomers, specifically Ins(1,4)$P_2$, Ins(1,3,4)$P_3$, and Ins(1,4,5)$P_3$, and Ins(1,3,4,5)$P_4$.

## Protocol 3

# Separation of inositol phosphates by HPLC

### Equipment and reagents

- Partisphere 5 SAX column 25 cm attached to a dual pump HPLC system
- On line radio-detector system, or a fraction collector for between 200–240 fractions and a static scintillation counter
- Elution buffer A: water
- Elution buffer B: 1 M ammonium phosphate pH 3.8 with phosphoric acid
- Scintillation cocktail
- Tritiated inositol phosphate standards

### Method

1. Wash the HPLC column extensively in water before each injection of sample. Neutralized extracts containing the water soluble inositol phosphates should be diluted with water (up to a volume of 2 ml) and injected onto the anion exchange column.
2. Bound radioactivity can be eluted with a linear gradient up to 100% B in water (A) at 1 ml/min over 120 min.

| Time (min) | %B |
|---|---|
| 0 | 0 |
| 5 | 0 |
| 100 | 100 |
| 120 | 100 |

The more negatively charged the inositol phosphate the tighter the binding to the strong anion exchange (SAX) column and the higher the concentration of ammonium phosphate required to elute it from the column. Thus, radioactive peaks expected to elute from the HPLC column are (in elution order) inositol (within the first 5 min), glycerophosphoinositol, InsPs, $InsP_2$s, $InsP_3$s ($InsP_3$ isomers should elute at around 40% B), $InsP_4$s, $InsP_5$s, and $InsP_6$.

3. The eluted radioactivity may be quantified using the on line detector. Alternatively 0.5 ml fractions can be collected and subject to static scintillation counting. Both methods require a suitable scintillation cocktail.
4. Identification of the inositol phosphates requires either internal standards or calibration of the HPLC separation with commercially available tritium-labelled external inositol phosphate standards (Amersham). Note that the elution of $InsP_3$ isomers can be 'marked' by inclusion of [$\gamma$-$^{32}$P]ATP, a compound readily available in most Biochemistry laboratories. For the internal standards [$^{14}$C]inositol phosphates can be obtained commercially or $^{32}$P-labelled inositol phosphates made in-house. Internal standards should be added to the samples just prior to injection and give an ideal method to identify inositol phosphate isomers based on the co-elution of the sample and the standard.

The above method gives one of the most reliable determinations of PLC activity by using HPLC to identify and quantify the production of [$^3$H]Ins(1,4,5)$P_3$ in cells. Several variations of the method exist, particularly at the acid quench and neutralization stage. This HPLC method is ideal for the analysis of [$^3$H]Ins(1,4,5)$P_3$ and Ins(1,3,4,5)$P_4$. It is beyond the scope of this review to detail the complex structural analysis of inositol phosphates but additional details on the analysis of inositides can be found in refs 1–3.

### 2.1.3 The Ins(1,4,5)$P_3$ mass assay

Although the above method can be readily achieved, it may not be universally applicable. For example, some primary cells will not tolerate the radioactivity required to label the inositol pool. A useful alternative to labelling is to use a mass assay method to quantitate Ins(1,4,5)$P_3$ (available from Amersham). This method makes use of the high quantities of Ins(1,4,5)$P_3$ binding protein found in adrenal gland cortical microsomes (obtain porcine adrenal glands from a local slaughterhouse) and is described in Protocol 4. In cells, the binding of Ins(1,4,5)$P_3$ to this protein causes the release of calcium from intracellular stores. Early research showed that this receptor had a high affinity for Ins(1,4,5)$P_3$ and was an ideal tool for a mass assay. The assay compares the displacement of authentic [$^3$H]Ins(1,4,5)$P_3$ by authentic unlabelled Ins(1,4,5)$P_3$ to that displaced by the sample over a concentration range of 0–100 nM.

## Protocol 4

## Mass measurement of Ins(1,4,5)$P_3$

### Reagents

- Fresh porcine adrenal glands
- Homogenization buffer: 20 mM $NaHCO_3$ pH 7.5, 1 mM DTT
- Incubation buffer: 100 mM Tris pH 9, 4 mM EDTA, 4 mM EGTA, 4 mg/ml BSA
- Ins(1,4,5)$P_3$ and tritiated Ins(1,4,5)$P_3$

### Method

1. Thoroughly homogenize fresh adrenal glands on ice in ice-cold homogenization buffer by blending (use four 15 sec bursts in a chilled Waring blender) followed by glass-Teflon homogenization.
2. Centrifuge the homogenate (5000 g, 15 min, 4 °C) and retain the supernatant. Centrifuge the supernatant (35 000 g, 20 min, 4 °C) and retain the pellet and supernatant (note that this supernatant can be used a source of cAMP binding protein, see below). Re-homogenize the pellet and re-centrifuge as before. Resuspend the pellet (approx. 40 mg/ml) and store at −80 °C in aliquots.

**Protocol 4** continued

3. Perchloric acid extract and neutralize the inositol phosphates from unlabelled cultures of cells or tissues essentially as described in Protocol 1. Approx. $1 \times 10^6$ cells are required per point.
4. Mix 25 μl of neutral extract, or the standard Ins(1,4,5)$P_3$ (concentration range 0–100 nM plus one point at 4 μM Ins(1,4,5)$P_3$ for non-specific binding), with 25 μl binding protein, the [$^3$H]Ins(1,4,5)$P_3$ (available from Amersham; use at least 5000 cpm per sample point in 25 μl), and 25 μl buffer, and incubate on ice for 30 min. Ideally, the dilution of the extract used should fall on the linear part of the standard curve obtained (i.e. equivalent to 50 nM).
5. Terminate the reaction by centrifugation in a microcentrifuge (12 000 g, 3 min, 4 °C) and aspirate the supernatant.
6. Dissolve the pellet in scintillant (add tissue dissolver as necessary) and determine the level of radioactivity.
7. The data is plotted as % B/Bo vs. [Ins(1,4,5)$P_3$] in nM.

$$\frac{\text{B (amount of [}^3\text{H]Ins(1,4,5)P}_3\text{ bound)}}{\text{Bo (Total [}^3\text{H]Ins(1,4,5)P}_3\text{ non-specific binding)}} \times 100$$

The mass of Ins(1,4,5)$P_3$ of the extracts can then be determined from the standard curve.

### 2.1.4 Ins(1,3,4,5)$P_4$ mass assay

In cells, some Ins(1,4,5)$P_3$ is converted to Ins(1,3,4,5)$P_4$, which also plays a role in calcium homeostasis. High affinity Ins(1,3,4,5)$P_4$ binding proteins have been described in brain. Rat cerebellum represents an ideal source of Ins(1,3,4,5)$P_4$ binding protein for use in a mass assay. Cerebella microsomes are prepared as described in Protocol 3. Incubations should proceed on ice at pH 5 in a suitable buffer with the standard Ins(1,3,4,5)$P_4$ (0–100 nM) and unknown samples, plus a fixed amount of [$^3$H]Ins(1,3,4,5)$P_4$ (5000 cpm). Terminate the reaction as described in Protocol 2 and quantify as described for the Ins(1,4,5)$P_3$ binding assay.

## 2.2 Protein kinase C assay

Protein kinase C is the major effector of the PLC signalling pathway. As such, determination of PKC activity can give a measure of this signalling pathway. Antibodies are available that can immunoprecipitate PKC isoforms and the activity of the protein kinase can then be assessed as described in Protocol 5.

## Protocol 5

## Measurement of PKC activity in immunoprecipitates

### Reagents

- Lysis buffer: 20 mM Hepes pH 7.4, 150 mM NaCl, 3 mM EDTA, 3 mM EGTA, 1 mM DTT, 0.5% (w/v) Nonidet P-40, 1 mM sodium orthovanadate, and protease inhibitor cocktail tablet (Boehringer Mannheim)
- Kinase buffer: 50 mM Hepes pH 7.4, 100 mM NaCl, 75 mM KCl, 10 mM $MgCl_2$, 2.5 mM $CaCl_2$, 1 mM sodium orthovanadate, 1 mM DTT, 10 mM β-glycerophosphate
- Lipid micelles: 500 μM phosphatidylserine, 100 μM diacylglycerol in 0.5% Triton X-100
- Anti-PKC antibody (e.g. Santa Cruz Biotechnology)
- Protein A (or G)-Sepharose (e.g. Sigma or Pharmacia)
- 1 mg/ml histone HIII (Sigma)
- 0.5% phosphoric acid

### Method

1. Prepare cell lysates in lysis buffer. It may be necessary to include additional Ser/Thr phosphatase inhibitors in this lysis buffer, such as 50 mM sodium fluoride and 2 μM microcystin-LR.
2. Incubate the lysates with an appropriate PKC antibody for 4 h at 4 °C. Immunoprecipitate with 20 μl protein A or protein G-Sepharose (50% slurry) for 2 h at 4 °C. A complete explanation of the theory and practice of immunoprecipitation and other immunological techniques can be found in ref. 4.
3. Wash the immunoprecipitates several times in lysis buffer.
4. To the immunoprecipitated PKC, add 5 μl of 1 mg/ml histone IIIS (or other suitable PKC acceptor peptide) in kinase buffer and 10 μl of lipid micelles. Start the reaction by adding 100 μM ATP containing 1 μCi [$\gamma$-$^{32}$P]ATP.
5. Incubate for 60 min at 30 °C.
6. If using histone as substrate, stop the reaction by adding SDS-PAGE sample buffer directly to the tube, mix, and heat to 100 °C for 5 min. Load the sample onto SDS-PAGE gels to separate the phosphorylated histone (a high percentage gel will be required such as 17.5% or a gradient gel 4-20%). Electrophorese the dye front off the bottom the gel to remove the excess ATP. Fix and stain the gels (see Section 6).
7. Dry the gel and visualize radioactive bands by phosphorimage analysis or autoradiography.
8. If an acceptor peptide is used, acidify the incubation with 20 μl of 5% acetic acid and spot a portion of the aqueous solution on to P81 phosphocellulose paper. The radioactive phosphorylated peptide is retained and the excess ATP is washed away following several changes of 0.5% phosphoric acid.
9. The bound radioactivity can be quantified by phosphorimage analysis or by scintillation counting.

**Notes**: PKC translocates to membranes following PLC-mediated DAG production and therefore SDS–PAGE and Western blot analysis of this phenomena (by analysis of PKC immunoreactivity in particulate and cytosolic fractions from lysates prepared in the absence of detergent) can be a useful measure of PKC activation. Additional information on PKC assays can be found in ref. 3.

## 2.3 Fluorescent quantification of changes in intracellular calcium concentration

The availability of fluorescent dyes that can detect changes in intracellular calcium concentration has permitted detailed analysis of calcium homeostasis in living cells. The dye of choice for calcium is the cell permeable acetoxymethyl ester of fura-2 (fura-2AM). A suitable method is described in Protocol 6.

## Protocol 6

## Measurement of intracellular calcium in cells in a fluorometer

### Reagents

- Loading buffer: 20 mM Hepes pH 7.4, 130 mM NaCl, 5 mM KCl, 1 mM $MgCl_2$, 1.5 mM $CaCl_2$, 10 mM glucose, 0.1% (w/v) BSA
- 1 mM fura-2 in DMSO

### Method

1. Cells should be incubated normal serum-free or in low serum medium for 48 h prior to treatment.
2. Non-adherent cells can be loaded with 4 μM fura-2AM in loading buffer for 30 min at 37 °C. Adherent cells should be detached from tissue culture-ware by trypsinization prior to fura-2AM loading, although this can lead to proteolysis of cell surface receptors. EDTA treatment can also be used to detach some cells (0.55 mM EDTA in PBS). Alternatively, cuvettes are available that can hold a small glass slide of adherent cells, enabling the measurement of calcium in the cell monolayer without the need to make a suspension of the cells.
3. Approx. $2 \times 10^6$ cells should be loaded into the fluorometer cuvette and fluorescence determined with dual excitation at 340 and 380 nm and emission at 510 nm. Such a system can then be used to assess the effect of extracellular calcium on the kinetics of intracellular calcium release. Furthermore, the effects of pharmacological agents known to interfere with Ins(1,4,5)$P_3$-mediated calcium release can be determined if cells are permeabilized.

Spatial aspects of intracellular calcium signalling can be visualized using appropriate confocal microscopes. In addition, these instruments can be calibrated to give reasonably accurate quantitation of calcium release. Note that additional information on the use specific fluorescent dyes to measure other signalling molecules (cAMP/NO) and other ions ($Na^+$) can be found in ref. 3 and the resource CD-ROM available from Molecular Probes (8).

## 3 Quantification of phospholipase D (PLD) activity in cells

Agonist stimulation of mammalian cells rapidly enhances phospholipase D-catalysed hydrolysis of the major plasma membrane phospholipid, phosphatidylcholine (PtdCho), to yield phosphatidic acid (PtdOH) and choline. Many agonists stimulate both PLD and PLC in cells and it was originally thought that the role of PLD was to potentiate PLC-initiated PKC activation by producing DAG from dephosphorylated PA. It is now clear the PtdCho-derived PA acts as a messenger in cells and is involved in regulating a diverse range of cellular processes, including secretion, the oxidative burst, and cystoskeleton reorganization.

Protocol 7 was developed for adherent fibroblast cell lines and may need adaptation to suit other systems. For example, the PtdCho pool in certain cells may need to be labelled with a particular fatty acid (e.g. palmitic acid or myristic acid or oleic acid) or other intermediates (e.g. LysoPAF). Assaying phospholipase D activity in cells uses the trans-phosphatidylation property of the enzyme that occurs during PtdCho hydrolysis. *In vivo*, the accepting nucleophile for the phosphatidyl moiety (after choline is released) is water, resulting in the generation of phosphatidic acid that is rapidly metabolized in cells to DAG. However, primary short chain alcohols are stronger nucleophilic acceptors in the PLD reaction and divert the phosphatidyl moiety to a metabolically more stable phosphatidylalcohol. Butan-1-ol is commonly used to generate phosphatidylbutanol since butan-2-ol is a good control for any non-specific alcohol effects. This method not only enables quantification of PLD activity in cells but prevents PLD-mediated signalling events by preventing the production of PtdOH without inhibiting PLD activity.

## Protocol 7

## Measurement of PLD activity in cells

### Equipment and reagents

- LK5DF TLC plates (Whatman) and suitable TLC tank
- Scintillation fluid and counter
- Labelling medium: normal growth medium with 0.5–1% fatty acid-free BSA containing 5 μCi/ml [$^3$H]palmitic acid (or other fatty acid such as myristate)[a]
- Butano-1-ol and butan-2-ol
- Post-labelling wash medium: DMEM containing 0.1% BSA, 0.0375% (w/v) sodium bicarbonate, 20 mM Hepes pH 7.4
- Ice-cold methanol
- Chloroform
- Solvent for TLC separation: the upper phase of 2,2,4 trimethylpentane/ethylacetate/acetic acid/$H_2O$ (50:110:20:100, by vol.)

**Protocol 7** continued

### Method

1 Grow cells to 70–80% confluence in normal growth medium (12- or 24-well plates will give sufficient material for analysis).

2 Remove the normal growth medium and replace with labelling medium. Culture the cells for 24–36 h to achieve near-equilibrium labelling of the PtdCho pool.

3 Prior to stimulation, remove the labelling medium and wash the cells. Pre-incubate in fresh wash medium containing 0.3% (v/v) butan-1-ol (or butan-2-ol as a control) for 10 min. Stimulate the cells at 37 °C with either fresh medium containing butan-1-ol or by diluting the ligand directly into the medium in the well (see Protocol 1).

4 To terminate the incubation, aspirate the medium and add 500 μl ice-cold methanol. Scrape the cell debris into a glass vial and wash the well with 200 μl methanol. Combine these two methanol washes.

5 Add 700 μl of chloroform to each vial. After 15 min, add 600 μl of water, vortex the samples, and centrifuge at 1000 g for 5 min to separate the phases.
**Note**: steps 3–5 are essentially a Bligh and Dyer-type extraction that results in the quantitative extraction of most lipids from the cell, and as such can be used to provide samples for DAG and ceramide mass analysis (see Protocol 8).

6 The lower organic phase should be washed with synthetic upper phase and then dried *in vacuo* in the bottom of a glass tube. Resuspend the dried lipids in 20–40 μl $CHCl_3$/MeOH (19:1, v/v) together with 10 μg phosphatidylbutanol standard (Avanti Polar Lipids, USA, or Lipid Products, UK) for spotting on LK5DF TLC plates.

7 Develop the TLC plates in separation solvent until the solvent front is 1–2 cm from the top of the plate.

8 The tritiated PLD product (phosphatidylbutanol) has an Rf of approx. 0.3–0.4 under these conditions and can be located by staining the plate in an iodine tank to locate the phosphatidylbutanol standard. Mark the area with a soft pencil and scrape the silica into a scintillation vial. Add scintillant and allow 24 h for the scintillant to release the radioactivity from the silica. Quantify labelled lipid by scintillation counting.

[a] Should the cells not tolerate low BSA, then the serum concentration should be lowered to 0.1–1%.

## 3.1 Measurement of DAG and ceramide mass

Mass measurement of diacylglycerol (DAG) utilizes a commercially available preparation of a bacterial lipid kinase that can phosphorylate DAG to produce phosphatidic acid, monoacylglycerol (MAG) to lysophosphatidic acid, and ceramide to ceramide phosphate. Under appropriate conditions, the assay described in Protocol 8 is quantitative, and can be used to mass assay cellular changes in total DAG (or MAG or ceramide). It is therefore a useful assay for studying both PLD- and PLC-mediated signalling events.

## Protocol 8

# Mass measurement of DAG (or ceramide) in cells

### Equipment and reagents

- Glass-backed silica $60F_{254}$ TLC plate $10 \times 10$ cm and TLC tank
- 300 μM phosphatidylserine in 0.6% Triton X-100
- 1 mM diacylglycerol and ceramide standards in chloroform/methanol (2:1, v/v)
- 100 mM DTT in water
- Kinase buffer: 250 mM imidazole pH 6.6, 250 mM NaCl, 62.5 mM $MgCl_2$, 5 mM EGTA
- *E. coli* DAG kinase (Calbiochem, UK)
- 0.5 mM ATP containing 1 μCi [$\gamma$-$^{32}$P]ATP in kinase buffer
- Stopping solution: $CHCl_3$/MeOH/ concentrated HCl (150:300:2, by vol.)
- Chloroform and methanol.
- TLC solvent: $CHCl_3$/MeOH/acetic acid (40:10:5, by vol.)

### Method

1. Prepare Bligh and Dyer-type extracts from treated and untreated cells as described in Protocol 7, steps 3–5. Dry the chloroform phase in the bottom of clean 2 ml glass tubes *in vacuo*.
2. Dry sufficient phosphatidylserine (PtdSer) in a glass test-tube under a stream of nitrogen and resuspend to give a final concentration of 300 μM (for 40 samples and five standards in duplicate a total of 3 ml PtdSer mixture is required).
3. Probe sonicate the PtdSer mixture and add 50 μl to each tube containing dried cellular lipids. Cap the tubes and sonicate the tubes in a bath sonicator for 30 min in ice to prevent evaporation.
4. To each tube, add 10 μl of 100 mM DTT, 20 μl kinase buffer, and 0.5–1 mU *E. coli* DAG kinase in 10 μl. Start the reaction by adding 10 μl of 0.5 mM ATP containing 1 μCi of [$\gamma$-$^{32}$P]ATP to each tube and incubate at 30 °C for 30–60 min.
5. Terminate the reaction by adding 1 ml of stopping solution. After 10 min, add 300 μl of $CHCl_3$ and 400 μl $H_2O$. Vortex the tubes and centrifuge at 1000 *g* for 5 min to split the phases. Discard the upper aqueous phase and dry the lower organic phase in the bottom of the tube *in vacuo*.
6. Resuspend the dried lipid in 20–40 μl of $CHCl_3$/ MeOH (19:1) and apply the sample near the bottom of a dry glass-backed silica $60F_{254}$ TLC plate as a 1 cm strip. Develop in $CHCl_3$/MeOH/acetic acid (40:10:5, by vol.) until the solvent front is close to the top of the plate (phosphatidic acid has a Rf of approx. 0.8 and ceramide phosphate Rf 0.4). Radiolabelled bands can be located by and quantified by phosphorimage analysis. The mass of phosphatidic acid produced is related to the mass of DAG by means of a standard curve using pure 1-stearoyl, 2-arachidonoylglycerol in the range 0–2000 pmol. If phosphorimage analysis is used, the original [$\gamma$-$^{32}$P]ATP will need to be quantified to demonstrate the efficiency of conversion and convert image units to pmol. A 1 in 50 dilution of the ATP stock is equivalent to 2000 pmol. If autoradiography is used, bands will need to be visualized on film and then the corresponding area of silica scraped off the plate and the level of radioactivity determined.

A complete guide to the measurement of phospholipid signalling, including phospholipase C, phospholipase D, phospholipase $A_2$, PI kinases, and sphingomyelin metabolites is given in ref. 2.

## 4 Phosphatidylinositol 3-kinase (PI 3-kinase) signalling in cells

Phosphoinositide (PI) 3-kinase activity has been implicated in a wide variety of cellular processes, including growth, secretion, and membrane ruffling. *In vivo*, PI 3-kinase can phosphorylate phosphatidylinositol, phosphatidylinositol phosphate, and PtdIns(4,5)$P_2$ to the appropriate 3-phosphorylated lipid. However, *in vivo* it is likely that the substrate specificity of PI 3-kinase is more restricted. The classical PI 3-kinase activity is a heterodimer consisting of a regulatory (p85) and a catalytic subunit (p110). The p85 subunit associates with cell surface receptors in a tyrosine phosphorylation-dependent manner via protein interaction domains, and ligation of these receptors stimulates lipid kinase activity.

Three methods are available to quantify this signal transduction pathway: PI 3-kinase activity can be quantified in immunoprecipitates (Protocol 9), by detection of one of the downstream targets of PI 3-kinase activation (phospho-PKB/Akt) (Protocol 10), or by measurement of 3-phosphorylated lipids in cells (Protocol 11). The last method is particularly applicable to PI 3-kinase isoforms which do not couple to receptors via p85. The role of PI 3-kinase in regulating cellular events can also be determined by assessing the effects of selective PI 3-kinase inhibitors such as Wortmannin (suggested dose 100 nM) and Ly294002 (suggested dose 10 $\mu$M).

## Protocol 9

### Measurement of PI 3-kinase activity in immunoprecipitates

#### Equipment and reagents

- Oxalate pre-treated LK5DF TLC plates: soak the TLC plate for 20 min in 2 mM potassium oxalate, 60% MeOH and air dry for 2 h
- Lysis buffer: 20 mM Hepes pH 7.4, 150 mM NaCl, 3 mM EDTA, 3 mM EGTA, 1 mM DTT, 0.5% (w/v) Nonidet P-40, 1 mM sodium orthovanadate, 50 mM NaF, 2 $\mu$M microcystin LR, and a protease inhibitor cocktail tablet (Boehringer Mannheim)
- Anti-p85 antibody (e.g. Serotec) or an anti-phosphotyrosine antibody (e.g. PY20 or 4G10)
- Protein A–Sepharose (or protein G–Sepharose)
- Kinase buffer: 25 mM Hepes pH 7.4, 5 mM $MgCl_2$
- Lipid micelles: dry sufficient PtdIns (10 mg/ml in chloroform/methanol, 1:1 stock) in a glass tube under a stream of nitrogen and sonicate to give 1.5 mg/ml PtdIns in 25 mM Hepes pH 7.5, 5 mM $MgCl_2$, and 0.5% sodium cholate
- Stop solution: $CHCl_3$/MeOH/$H_2O$ (5:10:2, by vol.)
- Chloroform
- 2.4 M HCl
- TLC buffer: $CHCl_3$/acetone/MeOH/acetic acid/$H_2O$ (40:15:15:12:8, by vol.)

**Protocol 9** continued

### Method

1. Lyse stimulated and unstimulated cells on ice in lysis buffer.
2. Immunoprecipitate PI 3-kinase activity using specific anti-p85 and protein A–Sepharose for 2–4 h at 4 °C (4). Wash the beads extensively with lysis buffer to ensure the immunocomplex is detergent-free. The efficiency of the assay is improved by performing the last wash in kinase buffer (without ATP) prior to commencing the reaction. The presence of 0.5 M lithium in the washes can also improve detection of immunoprecipitated PI 3-kinase activity. Remove all buffers by aspiration, leaving just the lipid kinase bound to the Sepharose beads.
3. Add 20 μl of the sonicated PtdIns per immunoprecipitation to the beads and warm to 37 °C.
4. Make a 5 μM ATP stock containing 5 μCi [$\gamma$-$^{32}$P]ATP per point in kinase buffer. Add 10 μl of the ATP to the immunoprecipitates (and lipids) and incubate at 37 °C for 60 min.
5. Terminate the reaction by adding 500 μl stop solution. Add 400 μl $CHCl_3$ and 150 μl of 2.4 M HCl, vortex, and split the phases by low speed centrifugation (1000 g, 5 min).
6. Transfer the lower organic phase to a glass tube and dry *in vacuo*. Resuspend the lipids in 20–40 μl $CHCl_3$/MeOH (19:1) and apply the samples to the lanes of the oxalate pre-treated LK5DF TLC plates. Develop the TLC plate in a tank pre-equilibrated with TLC buffer until the solvent front is close to the top of the plate.
7. Radioactivity can be visualized and quantified by phosphorimaging. For autoradiography, visualize the radioactive bands on film, scrape the corresponding area of silica from the plate into a tube, and count.
8. A suitable positive control of $^{32}$P-labelled PtdInsP can be generated from an anti-phosphotyrosine immunoprecipitate of PDGF-stimulated Rat-1 cells.

## 4.1 Measurement of PI 3-kinase downstream targets

The observation that PKB/Akt is phosphorylated in a PtdIns(3,4,5)$P_3$-dependent manner in cells provides a convenient assay for PI 3-kinase activation since phospho-specific PKB/Akt antibodies are now commercially available (New England Biolabs).

## Protocol 10

## Detection of phospho-PKB/Akt

### Equipment and reagents

- SDS–PAGE equipment and reagents
- Phospho-PKB/Akt antibody (New England Biolabs)
- Western blotting equipment and reagents
- Secondary antibody and enhanced chemiluminescence (Amersham)

**Protocol 10** continued

### Method

1. Lyse stimulated and unstimulated cells directly into SDS–PAGE sample buffer. Alternatively, a lysate can be prepared and then diluted into the SDS–PAGE buffer.
2. Use standard SDS-PAGE and Western blotting (4) using the phospho-PKB/Akt antibody at a starting dilution of 1:2000.
3. Visualize immunostaining with appropriate secondary antibody and enhanced chemiluminescence using autoradiography.

Note that as an alternative to this method, PKB/Akt can be immunoprecipitated and its phosphorylation of GSK3 assessed.

## 4.2 Quantification of 3-phosphorylated lipids in cells

Detection and quantification of 3-phosphorylated lipids in cells provides an unequivocal measure of changes in PI 3-kinase activity under stimulated conditions. However, this method requires considerable technical skill and large amounts of radioactivity. In brief, Protocol 11 describes how to label the lipids and perform the deacylation of the lipids prior to HPLC analysis. The method is sensitive and able to detect the very small quantities of 3-phosphorylated lipids produced in cells.

## Protocol 11

## Quantification of 3-phosphorylated inositol lipids in cells

### Equipment and reagents

- Partisphere 5 SAX column (25 cm) attached to dual pump HPLC system with fraction collector or on line radiodetector
- Phosphate-free growth medium, e.g. phosphate-free DMEM containing 0.1% BSA, 0.0375% (w/v) sodium bicarbonate, 20 mM Hepes pH 7.4
- 10 mCi [$^{32}$P]Pi in PBS (Amersham)
- Stopping solution: MeOH/1 M HCl (3:1, v/v)
- Chloroform
- 33% methylamine in methylated spirit/MeOH/$H_2O$/butan-1-ol (58:4:58:12, by vol.)
- Butan-1-ol/petroleum ether/ethyl formate (20:4:1, by vol.)
- HPLC elution buffer A: water
- HPLC elution buffer B: 0.5 M ammonium phosphate (pH 3.8 with phosphoric acid)

### Method

1. Grow cells to confluence in 10 cm dishes and quiesce in low serum-containing medium.
2. Wash the cells twice with a phosphate-free version of the medium.
3. Incubate the cells with phosphate-free medium containing 0.5–1 mCi/ml [$^{32}$P]Pi for 2 h at 37 °C.

**Protocol 11** continued

4 Wash the cells twice in the phosphate-free medium and then stimulate the cells as required.

5 Terminate incubations by adding 3 ml ice-cold stopping solution, and for adherent cells scrape into a 20 ml glass tube. (Non-adherent cells could be labelled in these large 20 ml tubes.)

6 Wash the dishes with stop solution and combine the two methanol washes.

7 Add 1 ml of chloroform and 20 μg of carrier lipid (Folch fraction; Sigma), vortex, and split the phases by adding 5 ml $CHCl_3$ and 1.25 ml of $H_2O$ (giving a Folch distribution of $CHCl_3/MeOH/H_2O$ - 8:4:3, by vol.). Following centrifugation (1000 g, 5 min), the lower organic phase should be washed twice with synthetic upper phase. The lower phase, containing the $^{32}P$-labelled lipids, can be stored at −20 °C at this stage.

8 Dry the lower phase *in vacuo* and resuspend in 5 ml of 33% methylamine in methylated spirit/MeOH/$H_2O$/butan-1-ol (58:4:58:12, by vol.) and incubate for 60 min at 53 °C with occasional vortexing.

9 Terminate the reaction by incubating the samples on ice and then dry *in vacuo*.

10 The glycerophosphoinositols can be separated from the fatty acid chains by adding 2 ml of $H_2O$ and 2.5 ml of butan-1-ol/petroleum ether/ethyl formate (20:4:1, by vol.). Vortex the samples and split the phases by centrifugation (1000 g, 5 min). The lower aqueous phase contains the glycerolinositol phosphates.

11 The glycerophosphoinositols can be resolved by HPLC on a 25 cm Partishere 5 SAX column eluted with a linear gradient of 0.5 M ammonium phosphate pH 3.8 in water over 120 ml at 1 ml/min (1). Collect fractions every 0.5 min and subject to liquid scintillation counting. This is essentially similar to the method described to resolve inositol phosphates (Protocol 1). However, a lower % of B is required since glycerophosphoinositols are less charged than their corresponding inositol phosphate. Although glycerophosphatidylinositol 4,5-bisphosphate contains the Ins(1,4,5)$P_3$ headgroup it will run on SAX HPLC closer to an authentic $InsP_2$ standard. Deacylated commercially available tritiated polyphosphoinositides (Amersham) can added be to the $^{32}P$-labelled samples as internal standards. This approach will give the added benefit of enabling the calculation of recovery through the processes, where a value of 75% or better is acceptable.

## 5 Measurement of small G protein function in cells

Small molecular weight GTPases of the Ras superfamily (SMGs; also known as p21 proteins) are ubiquitously expressed proteins involved in a wide variety of cellular processes, including growth, vesicle trafficking, membrane ruffling, cytoskeletal reorganization, regulation of PLD activity, regulation of the PI cycle, and nuclear transport. In resting cells, small molecular weight G proteins are bound to GDP. Upon activation, GDP is exchanged for GTP and SMGs assume an active conformation. The intrinsic GTP exchange and hydrolysis activities of SMGs are regulated by accessory proteins (guanine nucleotide exchange factors,

GEFs, and GTPase activator proteins, GAPs). Thus, increased GTP loading of a SMG is a reliable measure of activation. However, this assay requires large amounts of radioactivity and a suitable antibody to immunoprecipitate the small G protein of interest. Unfortunately, there are not suitable antibodies to all SMGs, although the anti-Ras antibody Y13-259 (Santa Cruz Biotechnology) is a notable exception and Protocol 12 details its use.

## Protocol 12

## Assessment of the guanine nucleotide status of Ras *in vivo*

### Equipment and reagents

- Phosphorimage or autoradiography facilities
- Polyethylenimine (PEI) TLC plates and suitable TLC tank
- Phosphate-free normal growth medium, e.g. DMEM containing 0.1% BSA, 0.0375% (w/v) sodium bicarbonate, 20 mM Hepes pH 7.4
- 10 mCi [$^{32}$P]Pi in PBS (Amersham)
- Anti-Ha-Ras agarose (Y13-259, Santa Cruz Biotechnology)
- Lysis buffer: 20 mM Hepes pH 7.4, 150 mM NaCl, 3 mM EDTA, 3 mM EGTA, 1 mM DTT, 0.5% (w/v) Nonidet P-40, 1 mM sodium orthovanadate, and a protease inhibitor cocktail tablet
- Elution buffer: 20 mM Tris–HCl pH 7.4, 10 mM EDTA, 2% SDS, 0.5 mM GDP, 0.5 mM GTP
- TLC buffer: 0.75 M $KHPO_4$ pH 3.4 with phosphoric acid (or 1 M LiCl)

### Method

1. Grow cells to confluence in 10 cm dishes and quiesce in low serum-containing medium.
2. Wash the cells twice with a phosphate-free medium.
3. Incubate the cells with phosphate-free medium containing 0.5–1 mCi/ml [$^{32}$P]Pi for 12–16 h at 37 °C.
4. Stimulate the cells as required and lyse in a suitable lysis buffer.
5. Immunoprecipitate Ras by adding anti-Ha-Ras agarose to the lysates. Harvest the beads by centrifugation, discard the supernatant, and wash the beads extensively with lysis buffer.
6. Release the bound $^{32}$P-labelled guanine nucleotides by washing the beads in elution buffer for 20 min at 68 °C. Spot the eluate onto PEI TLC plates and develop in 0.75 M $KHPO_4$ pH 3.4 with phosphoric acid (or 1 M LiCl). A lane on the TLC plate should be devoted to standards. In general, the GTP will not move far from the origin whereas the GDP will move considerably further.
7. Visualize and quantify the radioactivity by phosphorimage analysis.
8. The amount of loading is best expressed as follows:

$$\frac{\text{Amount of radioactivity in GTP}}{\text{GDP radioactivity} + \text{GTP radioactivity}} \times 100$$

This equation takes into account the extra phosphate on GTP compared to GDP.

## 5.1 Additional notes on the measurement of SMG protein activation

Protocol 12 relies on the relatively low rate of GTP hydrolysis of Ras proteins. For other small G proteins, this may not be the case and therefore trapping the GTP bound form before hydrolysis occurs may not be practicable. Immunoprecipitating antibodies are not available for other Ras superfamily members (such as Arf, the Rho family, Rabs, Rans, etc.), a problem that can be overcome if these proteins are linked to a suitable purification tag such as the HA, Myc, or FLAG epitopes.

At endogenous levels of expression seen in cells, the amount of GTP-loaded Ras may only double following stimulation. An elegant method for the detection of SMG activation has been developed recently following the identification and cloning of various effector proteins that bind the active (GTP bound) form of certain SMGs (5). The domain of the effector protein responsible for the interaction with the SMG is expressed recombinantly with a purification tag (usually glutathione *S*-transferase—GST) and is used to 'pull-down' the active SMG from cell lysates. Thus, the Ras binding domain of Raf1, the Rac1/Cdc42 binding domain of Pak1, and the Rho binding domain of Rhotekin are all suitable for this assay. Protocol 13 describes a generalized example of the cloning strategy required to generate Rho family member effector binding domains suitable for the assay described.

### Protocol 13

### Example method for the determination of the activation status of other SMG proteins

1. The Cdc42/Rac2 binding domain of PAK1 encompasses amino acids 45–141. The Rho binding domain of Rhotekin consists of the N terminal 90 amino acids. These constructs are cloned into a vector system to generate GST-fusion proteins and expressed in *E. coli* at 30 °C. 1 litre of bacterial culture generates milligram quantities of a GST-fusion protein under ideal conditions.
2. Following induction of expression, bacteria are harvested by centrifugation, lysed in detergent, and clarified by centrifugation. Glutathione–Sepharose is then used to affinity purify the GST-fusion protein. The resin is washed extensively to remove any contaminating proteins. A small aliquot of the beads is removed and separated by SDS–PAGE to assess the presence and purity of the of the GST-fusion protein. Note that the GST-fusion protein is not eluted from the beads with the glutathione.
3. A small amount (equivalent to 1–2 μg of fusion protein) of the immobilized SMG binding domain is then added to cell lysates and, following an incubation of a few hours, the beads are harvested by microcentrifugation and the supernatant discarded.
4. The bound proteins are released by incubation of the beads with SDS–PAGE sample buffer and analysed for the presence of the small G protein by SDS–PAGE and Western blotting (4). Several suppliers sell SMG antibodies suitable for blotting. Increased amounts of the SMG protein associated with the domains used indicates increased activation of the G proteins.

## 6 Measurement of MAP kinase activation

The MAP kinases are a family of serine/threonine protein kinases activated by dual phosphorylation. MAP kinases can be broadly divided into two groups: the *e*xtracellular signal *r*egulated *k*inase (ERK MAP kinase pathway) and the stress-activated protein kinases (consisting of the JNK pathway and the p38 MAP kinase pathway). Stimulation of cells with mitogens such as PDGF results in activation of Ras that initiates the ERK kinase cascade. The final stage of this cascade results in the phosphorylation of the ERK MAP kinases termed p42 and p44 and stimulation of gene transcription (e.g. of the Elk-1 gene). Cellular stresses such as UV irradiation or heat shock activate the JNK and/or the p38 kinase cascade. Activation of the JNK cascade ultimately results in phosphorylation of the Jun transcription factor by the Jun N terminal kinase (JNK). Cellular stresses and cytokines may also stimulate the p38 kinase cascade via activation of SMGs, ultimately resulting in increased phosphorylation of MAPKAPK2/3 by p38. For these kinase cascades, quantification of the activity of the last kinase in the cascade represents the most convenient read-out for stimulation of each pathway.

### 6.1 ERK kinase activity

Mitogens stimulate both p42 and p44 ERK by increasing the phosphorylation of threonine and tyrosine residues in characteristic Thr–Glu–Tyr motifs in a reaction catalysed by MAP kinase kinase (MEK1). The signal-regulated tyrosine phosphorylation of the p42/p44 ERK kinases was originally observed during anti-phosphotyrosine immunoblotting, as the p42/p44 bands appeared to alter electrophoretic mobility. This observation was developed into a 'band shift' assay for p42/p44. However, more convenient and quantitative methodologies are now available. The availability of suitable antibodies (Santa Cruz Biotechnology, New England Biolabs) permits the simple detection of the activated forms of these proteins using a method similar to that described in Protocol 14.

**Protocol 14**

### Detection of phosphorylated MAP kinases by Western blotting

#### Equipment and reagents

- Autoradiography facilities
- SDS–PAGE gel kit, blotting apparatus
- Enhanced chemiluminescence kit (ECL Amersham)
- Phospho-specific antibody for ERK1, ERK2, JNK, and p38
- Lysis buffer: 20 mM Hepes pH 7.4, 150 mM NaCl, 3 mM EDTA, 3 mM EGTA, 1 mM DTT, 0.5% (w/v) Nonidet P-40, 1 mM sodium orthovanadate, 50 mM NaF, 2 μM microcystin LR, and a protease inhibitor cocktail tablet

**Protocol 14** continued

### Method

1 Lyse stimulated and unstimulated cells on ice in SDS–PAGE sample buffer or using the lysis buffer and microcentrifuge (12 000 g, 10 min, 4 °C) to remove insoluble material.

2 Use a standard SDS–PAGE and Western blotting protocol and blot using the phospho-specific antibody of choice. Note that 5% BSA is preferred to 5% non-fat milk powder as the blocking agent when studying tyrosine phosphorylation.

3 Visualize using ECL and autoradiography.

Protocol 15 relies on immunoprecipitation of the enzyme followed by incubation with a suitable protein substrate and quantification of its phosphorylation. For p42/p44 ERK kinases the substrate of choice is myelin basic protein (MBP), for JNK the substrate is the N terminal portion of c-Jun (which is commercially available as a GST-fusion protein), and for p38 use ATF2 as substrate (also commercially available as a GST-fusion protein).

## Protocol 15

## Quantitation of MAP kinase activity

### Equipment and reagents

- SDS–PAGE gel kit
- Lysis buffer (see Protocol 14)
- Protein A–Sepharose (or protein G–Sepharose)
- Protein stain: 0.125% (w/v) Coomassie R-250, 50% ethanol, 10% acetic acid
- Kinase buffer: 20 mM Hepes pH 7.5, 20 mM β-glycerophosphate, 10 mM $MgCl_2$, 10 mM $MnCl_2$, 1 mM DTT, 50 μM sodium orthovanadate
- Destain I: 50% ethanol, 10% acetic acid
- Destain II: 10% ethanol, 5% acetic acid

### Method

1 Lyse stimulated and unstimulated cells on ice in lysis buffer and centrifuge in a microcentrifuge (12 000 g, 10 min, 4 °C) to remove insoluble material.

2 Add the appropriate antibody to the cleared lysates and mix for 1–2 h at 4 °C. Capture the complexes by adding 20 μl of protein A–Sepharose (50% slurry) for 1–2 h, 4 °C.

3 Wash the immunoprecipitates once with lysis buffer at 4 °C and twice with kinase buffer at 4 °C.

4 Resuspend the beads in kinase buffer containing 2 μg of myelin basic protein (or GST-Jun or GST-ATF2 as appropriate), 10 μM ATP containing 1 μCi [$\gamma$-$^{32}$P]ATP, and incubate at 37 °C.

**Protocol 15** continued

5. Terminate the incubation by adding SDS-PAGE sample buffer, vortex well, and incubate at 100 °C for 5 min.
6. Load the samples onto SDS-PAGE gels and electrophorese. It is important to run the dye front from the gels to remove the excess [$^{32}$P]ATP.
7. Fix and stain the gels using a standard Coomassie-type protocol. The processes of fixing and staining gels with Coomassie can be accelerated with no loss in quality by warming the separate washes in a microwave. Make up Coomassie protein stain and microwave the gel in stain on a medium-high setting for 2 min. Avoid boiling the stain. Wait for 15 min and remove the stain. Replace with destain I and microwave for 2 min on a medium-high setting. Wait 15 min, replace destain I with destain II, and microwave on a medium-high setting for 2 min.
8. Dry the gels prior to quantifying the radioactive bands by phosphorimage analysis.

## 7 cAMP signalling in cells

cAMP represents the most thoroughly characterized messenger produced in cells in response to stimulation of G protein-coupled hormone receptors. cAMP is produced from ATP by the action of adenylate cylcase and mediates many of its actions by binding to the regulatory subunit of protein kinase A. The PKA regulatory subunit can be used as an effective mass assay for cAMP production in cells and a crude preparation is readily obtained from the cytosolic fraction of adrenal glands. Thus the reader should refer to Protocol 4, where the particulate fraction of adrenal glands was used to prepare Ins(1,4,5)$P_3$ binding protein. A brief summary of the cAMP mass assay is given in Protocol 16. Additional information on this and complementary assays used to measure adenylate cyclase activity can be found in refs 3 and 6.

### Protocol 16

## Measuring cAMP mass in cultured cells[a]

### Equipment and reagents

- 25 mm filter discs (Whatman GF/B)
- Vacuum manifold and pump
- Hank's-buffered saline containing 10 mM glucose, 0.1% BSA, and 25 mM Hepes pH 7.5
- Stopping solution: 1.75% perchloric acid
- Neutralizing solution: 1.2 M $KHCO_3$
- cAMP assay buffer: 0.1 M Tris–HCl pH 7.4, 20 mM EDTA, 400 mM NaCl, 20 mM theophylline (warm this buffer to fully dissolve the reagents)
- Ice-cold wash buffer: 20 mM Tris–HCl pH 7.5, 5 mM EDTA

**Protocol 16** continued

### Method

1 Obtain the cytosol from homogenized adrenal cortex and store in aliquots at −80 °C (see Protocol 4).

2 Culture cells in 24-well plates in normal medium to confluence and quiesce for 24–48 h in a serum-free version of normal growth medium.

3 Wash the cells in Hank's-buffered saline containing 10 mM glucose, 0.1% BSA, and 25 mM Hepes pH 7.5, and stimulate with the required agonist at 37 °C in the same medium.

4 To quench the reaction, aspirate the medium, add 400 μl of 1.75% perchloric acid, and stand on ice for 15 min.

5 Scrape the cells into a microcentrifuge tube, neutralize with 100 μl of 1.2 M $KHCO_3$, and centrifuge (2000 g, 10 min, 4 °C). The neutral extracts can be stored at −20 °C. It is important to perform a control extract and neutralization on the cAMP standard.

6 In a 5 ml tube, mix 25 μl from one sample, or a cAMP standard (concentration range; 100 μM–100 fM), with 25 μl of cAMP assay buffer, and 0.25 μCi [$^3$H]cAMP.

7 Add the cAMP binding protein (25–50 μl), mix thoroughly, and incubate at room temperature for 60 min.

8 Terminate the reaction by adding 3 ml of ice-cold wash buffer to each tube and rapidly filtering each sample through separate Whatman GF/B 25 mm filter discs (a vacuum manifold capable of 25 samples at a time and a suitable pump is required to process large numbers of samples effectively).

9 Wash the filters with twice with 3 ml of ice-cold wash buffer.

10 Quantify the radioactivity bound to the discs by scintillation counting.

[a] Note that cAMP radioimmunoassay kits are commercially available (suppliers include Amersham, Dupont NEN).

## 8 Changes in cellular localization during signalling

It has become increasingly apparent that signal transduction processes cause key enzymes to change location in cells. For example, stimulation of cells with growth factors causes phospholipase C, PI 3-kinase, and PKC to translocate to the plasma membrane. Small G proteins also translocate from apparently cytosolic locations to the particulate fraction of stimulated cells and, as such, translocation can act as a competent marker of signal transduction pathway activation. Thus, Western blotting analysis of the redistribution of proteins between membrane and cytosolic fractions prepared from cell homogenates can be used to assess activation. However, not all proteins can be so conveniently analysed. For example, PLD translocates between two membrane locations during stimulation, and activated ERK and JNK MAP kinases translocate to the nucleus. Thus, methods that can relate the exact subcellular spatial and temporal organization of signal transduction events are becoming increasingly valuable. However, caveats remain. The expression levels of signal transduction enzymes are often so low that visualization of endogenous protein cannot be achieved. Therefore, localization studies of signal transduction proteins frequently rely on transfection of cells and overexpression of the exogenous protein fused to an appropriate epitope tag. Visualization of the protein of interest requires that the cells are fixed and the location of the tagged protein revealed with an appropriate antibody. Protocol 17 describes one way to investigate subcellular localization of a protein tagged with the haemagluttanin (HA) epitope (7).

## Protocol 17

### A general protocol for the visualization of HA-tagged proteins in fixed cells

#### Equipment and reagents

- Tissue culture equipment and reagents
- Transfection equipment and reagents
- PBS, paraformaldehyde, acetone
- Glycerol, DABCO

#### Method

1. Grow cells on adherent coverslips, suitable for mounting in a microscope, in normal growth medium. Transfect subconfluent cells with a mammalian expression vector containing the DNA of the protein of interest fused to an appropriate tag (such as HA). Transfection methods are essentially dependent upon the cell type being studied. Although the traditional calcium phosphate methods and electroporation are effective, the method of choice often utilizes cationic lipids in proprietary preparations (Fugene, Lipofectamine, DOTAP, DOPE, DOTMA). For the cationic lipid methods (except Fugene), it is important to transfect the cells in the absence of serum, since serum can reduce transfection efficiency. The amount of DNA used in the transfection should be titrated to the lowest amount that gives an acceptable signal for detection. The endogenous promoter for the protein of interest could be used to regulate expression from the vector to provide close to normal levels of expression.
2. Incubate the DNA and transfection agent with cells for 2 h and then replace the transfection medium with normal growth medium containing serum to allow the cells to recover for several hours (anywhere from 6–48 h, according to the efficiency of expression and stimulate as required).
3. To visualize the expressed proteins, wash the cells in PBS and fix in 4% paraformaldehyde for 5–10 min. Wash the slides in PBS and permeabilize by immersion in acetone at −20 °C for 5–10 min. Fixation and permeabilization should be determined empirically. Note that detergents such as 0.5% (w/v) CHAPS can be a more gentle permeabilization method.

**Protocol 17** continued

4. The commercially available anti-HA antibody '12CA5' can be used to visualize HA-fusion proteins. Incubate the fixed cells in PBS containing 20% heat-inactivated goat serum and the 12CA5 anti-HA antibody at a dilution of 1:100 for 1 h.
5. Wash the cells in PBS and add Texas Red conjugated anti-IgG antibody (Southern Biotechnology) for 1 h in PBS containing 20% heat-inactivated goat serum.
6. Wash the slides in cold PBS.
7. Dry the slides, mount in glycerol/DABCO, and visualize using confocal microscopy (excitation at 596 nm and emission at 615 nm for Texas Red). Note that proprietary reagents such as Prolong (Molecular Probes) can be used in place of DABCO to mount the slides.

## 8.1 Changes in cytoskeletal architecture

Activation of phospholipases, PI 3-kinase, and small G proteins stimulates dramatic rearrangement of the cells cytoskeleton. In particular, the actin cytoskeleton can undergo rapid and dramatic cycles of polymerization/depolymerization demonstrating the cells ability to alter shape for movement, secretion, or mitogenesis. Thus, visualization of cytoskeletal changes can be a valuable readout of signal transduction events. Protocol 18 describes the use of rhodamine-conjugated phalloidin to specifically label endogenous filamentous actin.

## Protocol 18

## Visualizing changes in the actin cytoskeleton

### Equipment and reagents

- Suitable fluorescence microscope, preferably confocal
- 0.33 μM rhodamine-conjugated phalloidin pH 7.5
- Wash buffer: Hank's-buffered saline containing 10 mM glucose, 0.1% BSA, and 25 mM Hepes

### Method

1. Grow cells on adherent coverslips, suitable for mounting in a microscope, in normal growth medium to subconfluency (30–40%).
2. Incubate the cells in serum-free medium for 24–48 h to induce quiescence.
3. Wash the cells in wash buffer and stimulate with agonists in this medium as required.
4. To visualize polymerized actin, wash the cells in ice-cold PBS, fix in 4% paraformaldehyde for 5–10 min (determine empirically), wash in PBS, and permeabilize by immersion in acetone at −20 °C for 5–10 min.
5. Incubate the cells in 0.33 μM rhodamine-conjugated phalloidin for 30 min.
6. Wash the slides in PBS and mount in glycerol/DABCO or Prolong. Visualize actin filaments by fluorescence microscopy.

## 8.2 Visualization of signal transduction events in living cells

An attractive possibility for studying signal transduction is the emerging technique of studying these events in living cells. This approach relies on the availability of either a confocal laser microscope or a deconvoluting microscope with digital camera. In either case, a heated and gassed stage is required to maintain the cells throughout the course of the experiment. The discovery and molecular cloning of the intrinsically fluorescent protein, green fluorescent protein (GFP), allows any protein to be fluorescently labelled using molecular cloning techniques. There are several important considerations when using GFP to analyse the location of signal transduction proteins. GFP is a 27 kDa protein and it is important to demonstrate that the presence of the tag does not interfere with the biochemical function(s) of the protein of interest. Equally, the tag itself should not influence the observed location (this can be investigated by comparing the location of the protein of interest when tagged with GFP-tagged protein to that when it is HA-tagged). Finally, the GFP reaction requires molecular oxygen so it is important to maintain the cells in a gassed environment. Additionally, if the pH of the medium should drop below pH 7, the fluorescence of GFP will reduce. Nevertheless, GFP technology presents a powerful tool to study the location of proteins in living cells.

Recently, protein modules have been identified that specifically bind second messenger molecules. For example, certain pleckstrin homology (PH) domains bind specifically and with high affinity to the lipid messenger PtdIns(3,4,5)$P_3$. The GFP-tagged version of a PH domain generates a 'biosensor' which, when transfected into cells, can be viewed in real-time translocating to membranes in a PI 3-kinase-dependent manner.

The protocols outlined in this chapter provide convenient and reproducible methods for assaying a wide variety of signal transduction pathways. However, not everything can be covered in the scope of this review. There are many books dedicated to techniques suitable for studying most aspects of signal transduction and some of these are listed in the references.

# References


1. Irvine, R. F. (ed.) (1990). *Methods in inositide research*. Raven Press, New York.
2. Bird, I. A. (ed.) (1998). *Phospholipid signalling protocols*. Methods in Molecular Biology series, Vol. 105. Humana Press, New Jersey.
3. Kendall, D. A. and Hill, S. J. (ed.) (1995). *Signal transduction protocols*. Methods in Molecular Biology series, Vol. 41. Humana Press, New Jersey.
4. Harlow, E. and Lane, D. (1988). *Antibodies: a laboratory manual*. Cold Spring Harbor Laboratory.
5. Sander, E. E., ten Klooster, J. P., van Delft, S., van der Kammen, R. A., and Collard, J. G. (1999). *J. Cell Biol.*, **147**, 1009.
6. Milligan, G. (ed.) (1992). *Signal transduction: a practical approach*. IRL Press, Oxford.
7. Beesley, J. E. (ed.) (1993). *Immunocytochemistry: a practical approach*. IRL Press, Oxford.
8. Haugland, R. P. *Handbook of fluorescent probes and research chemicals*. Available as a CD-ROM from Molecular Probes, Eugene, OR, USA.

# List of suppliers

**Ambion, Inc.**, 2130 Woodward Street, Austin, TX 78744-1832, USA
Tel: 512 651 0200
Fax: 512 651 0201

**Amersham Pharmacia Biotech UK Ltd**, Amersham Place, Little Chalfont, Buckinghamshire HP7 9NA, UK (see also Nycomed Amersham Imaging UK; Pharmacia)
Tel: 0800 515313
Fax: 0800 616927
URL: http//www.apbiotech.com/

**Anderman and Co. Ltd**, 145 London Road, Kingston-upon-Thames, Surrey KT2 6NH, UK
Tel: 0181 5410035
Fax: 0181 5410623

**BAbCO** (Berkeley Antibodies Company), 1223 South 47th Street, Richmond, CA 94804, USA

**Beckman Coulter (UK) Ltd**, Oakley Court, Kingsmead Business Park, London Road, High Wycombe, Buckinghamshire HP11 1JU, UK
Tel: 01494 441181
Fax: 01494 447558
URL: http://www.beckman.com/
Beckman Coulter Inc., 4300 N. Harbor Boulevard, PO Box 3100, Fullerton, CA 92834-3100, USA
Tel: 001 714 8714848
Fax: 001 714 7738283
URL: http://www.beckman.com/

**Becton Dickinson and Co.**, 21 Between Towns Road, Cowley, Oxford OX4 3LY, UK
Tel: 01865 748844 Fax: 01865 781627
URL: http://www.bd.com/
Becton Dickinson and Co., 1 Becton Drive, Franklin Lakes, NJ 07417-1883, USA
Tel: 001 201 8476800
URL: http://www.bd.com/

**Bio 101 Inc.**, c/o Anachem Ltd, Anachem House, 20 Charles Street, Luton, Bedfordshire LU2 0EB, UK
Tel: 01582 456666
Fax: 01582 391768
URL: http://www.anachem.co.uk/
Bio 101 Inc., PO Box 2284, La Jolla, CA 92038-2284, USA
Tel: 001 760 5987299
Fax: 001 760 5980116
URL: http://www.bio101.com/

**Bio-Rad Laboratories Ltd**, Bio-Rad House, Maylands Avenue, Hemel Hempstead, Hertfordshire HP2 7TD, UK
Tel: 0181 3282000
Fax: 0181 3282550
URL: http://www.bio-rad.com/
Bio-Rad Laboratories Ltd, Division Headquarters, 1000 Alfred Noble Drive, Hercules, CA 94547, USA
Tel: 001 510 7247000
Fax: 001 510 7415817
URL: http://www.bio-rad.com/

**Boehringer Mannheim** (see Roche)

**Bracco Diagnostics**, PO Box 5225, Princeton, NJ 08543, USA

**Brandel**, 8561 Atlas Drive, Gaithersburg, MA 20877, USA
Brandel, c/o SEMAT Technical (UK) Ltd, One Executive Park, Hatfield Road, St. Albans, Hertfordshire AL1 4TA, UK

**Calbiochem-Novabiochem Corp.**, CN Biosciences (UK) Ltd, Boulevard Industrial Park, Padge Road, Beeston, Nottingham NG9 2JR, UK
Tel: 0115 943 0840
Freefone: 0800 622935
Fax: 0115 943 0951
Calbiochem-Novabiochem Corp., 10394 Pacific Center Court, San Diego, CA 92121, USA
Calbiochem-Novabiochem Corp., PO Box 12087, La Jolla, CA 92039-2087, USA
Tel: 858 450 9600 Fax: 858 453 3552
URL: http//:www.calbiochem.com/

**Caltag Laboratories**, TCS Biologicals, Ltd, Botolph Claydon, Buckingham, MK18 2LR, UK
Tel: 01296 714555
Fax: 01296 714806
Caltag Laboratories, 1849 Old Bayshore Hwy., Suite 200, Burlingame, CA 94010, USA
Toll Free: 800 874 4007
Tel: 650 652 0468 Fax: 650 652 9030

**CellPoint Scientific**, 9210 Corporate Blvd. 220, Rockville, MD 20850, USA
Tel: 301 590 0055
URL: http//:www.roboz.com/

**Chroma Technology Corp.**, 72 Cotton Mill Hill, Unit A9, Brattleboro, VT 05301, USA

**CP Instrument Co. Ltd**, PO Box 22, Bishop Stortford, Hertfordshire CM23 3DX, UK
Tel: 01279 757711
Fax: 01279 755785
URL: http//:www.cpinstrument.co.uk/

**Dupont (UK) Ltd**, Industrial Products Division, Wedgwood Way, Stevenage, Hertfordshire SG1 4QN, UK
Tel: 01438 734000
Fax: 01438 734382
URL: http://www.dupont.com/
Dupont Co. (Biotechnology Systems Division), PO Box 80024, Wilmington, DE 19880-002, USA
Tel: 001 302 7741000
Fax: 001 302 7747321
URL: http://www.dupont.com/

**Eastman Chemical Co.**, 100 North Eastman Road, PO Box 511, Kingsport, TN 37662-5075, USA
Tel: 001 423 2292000
URL: http//:www.eastman.com/

**Electron Microscopy Sciences**, 480 S. Democrat Road, Gibbstown, NJ 08027, USA
Tel: 800 222 0342
Fax: 609 423 4389Electron Microscopy Services

**Eppendorf**, Brinkmann Instruments, Inc., One Cantiague Road, PO Box 1019, Westbury, NY 11590-0207, USA
Tel: 1 800 645 3050; 516 334 7500
Fax: 1 516 334 7506

**Fisher Scientific UK Ltd**, Bishop Meadow Road, Loughborough, Leicestershire LE11 5RG, UK
Tel: 01509 231166
Fax: 01509 231893
URL: http://www.fisher.co.uk/

Fisher Scientific, Fisher Research, 2761 Walnut Avenue, Tustin, CA 92780, USA
Tel: 001 714 6694600
Fax: 001 714 6691613
URL: http://www.fishersci.com/

**Fluka**, PO Box 2060, Milwaukee, WI 53201, USA
Tel: 001 414 2735013
Fax: 001 414 2734979
URL: http://www.sigma-aldrich.com/
Fluka Chemical Co. Ltd, PO Box 260, CH-9471 Buchs, Switzerland
Tel: 0041 81 7452828
Fax: 0041 81 7565449
URL: http://www.sigma-aldrich.com/

**Gibco-BRL** (see Life Technologies)

**GraphPad Software**, 5755 Oberlin Drive, No. 110, San Diego, CA 92121, USA

**Hybaid Ltd**, Action Court, Ashford Road, Ashford, Middlesex TW15 1XB, UK
Tel: 01784 425000
Fax: 01784 248085
URL: http://www.hybaid.com/
Hybaid US, 8 East Forge Parkway, Franklin, MA 02038, USA
Tel: 001 508 5416918
Fax: 001 508 5413041
URL: http://www.hybaid.com/

**HyClone Laboratories**, 1725 South HyClone Road, Logan, UT 84321, USA
Tel: 001 435 7534584
Fax: 001 435 7534589
URL: http//:www.hyclone.com/

**Invitrogen Corp.**, 1600 Faraday Avenue, Carlsbad, CA 92008, USA
Tel: 001 760 6037200
Fax: 001 760 6037201
URL: http://www.invitrogen.com/
Invitrogen BV, PO Box 2312, 9704-CH Groningen, The Netherlands
Tel: 00800 53455345
Fax: 00800 78907890
URL: http://www.invitrogen.com/

**Jackson ImmunoResearch Laboratories, Inc.**, Stratech Scientific Ltd, 61-63 Dudley Street, Luton, Bedfordshire LU2 0NP, UK
Tel: 01582 529000
Fax: 01582 481895
Jackson ImmunoResearch Laboratories, Inc., 872 West Baltimore Pike, PO Box 9, West Grove, PA 19390, USA
Tel: 800 367 5296, 610 869 4024
Fax: 610 869 0171

**Kinematica AG**, Luzernerstrasse 147a, CH-6014 Littau-Lucerne, Switzerland
Distributed in the UK by: Philip Harris Scientific, 618 Western Avenue, Park Royal, London W3 0TE, UK

**Levington**, The Scotts Company UK Ltd, Paper Mill Lane, Bramford, Ipswich IP8 4BZ, UK

**Life Technologies Ltd**, PO Box 35, 3 Free Fountain Drive, Inchinnan Business Park, Paisley PA4 9RF, UK
Tel: 0800 269210
Fax: 0800 243485
URL: http://www.lifetech.com/
Life Technologies Inc., 9800 Medical Center Drive, Rockville, MD 20850, USA
Tel: 001 301 6108000
URL: http://www.lifetech.com/

**Matek Corp.**, Homer Avenue, Ashland, MA 01721, USA
Tel: 508 881 6771; 800 634 9018
Fax: 508 879 1532

**Merck Ltd**, Hunter Boulevard, Magna Park, Lutterworth LE17 4XN, UK
Tel: 0800 22 33 44
Fax: 01455 55 85 86
Merck Ltd, Merck House, Poole BH15 1TD, UK
Tel: +44 (0) 1202 669 700
Fax: +44 (0) 1202 665 599

**Merck Sharp & Dohme Research Laboratories**, Neuroscience Research Centre, Terlings Park, Harlow, Essex CM20 2QR, UK
URL: http://www.msd-nrc.co.uk/
MSD Sharp and Dohme GmbH, Lindenplatz 1, D-85540, Haar, Germany
URL: http://www.msd-deutschland.com/

**Midland Certified Reagent Company**, 3112-A West Cuthbert Avenue, Midland, TX 79701, USA
Tel: 800 247 8766
Fax: 915 694 2387

**Millipore (UK) Ltd**, The Boulevard, Blackmoor Lane, Watford, Hertfordshire WD1 8YW, UK
Tel: 01923 816375
Fax: 01923 818297
URL: http://www.millipore.com/local/UKhtm/
Millipore Corp., 80 Ashby Road, Bedford, MA 01730, USA
Tel: 001 800 6455476
Fax: 001 800 6455439
URL: http://www.millipore.com/

**Minnesota Molecular**, 5129 Morgan Avenue South, Minneapolis, MN 55419, USA
Tel: 800 990 6268
Fax: 612 920 6590

**Molecular Probes, Inc.**, Cambridge Bioscience, 24-25 Signet Court, Newmarket Road, Cambridge CB5 8LA, UK
Tel: +44 1 223 316855
Fax: +44 1 223 360732
Molecular Probes, PO Box 22010, Eugene, OR 97402-0469, 4849 Pitchford Ave., Eugene, OR 97402-9165, USA

**MSE**, owned by Sanyo Gallenkamp

**Nalge Nunc International**, 75 Panorama Creek Drive, Rochester, NY 14625, USA
Tel: 716 586 8800
Nalge Nunc International, Foxwood Court, Rotherwas Industrial Estate, Hereford HR2 6JQ, UK
Tel: 011 44 1432 263933
Fax: 011 44 1432 351923

**New England Biolabs**, 32 Tozer Road, Beverley, MA 01915-5510, USA
Tel: 001 978 9275054

**New Brunswick Scientific Co. Inc.**, PO Box 4005, 44 Talmadge Road, Edison, NJ 08818-4005, USA
Tel: 1 732 287 1200
Fax: 1 732 287 4222
URL: http://www.nbsc.com/

**Nickerson-Zwaan Ltd**, Rothwell, Market Rasen, Lincoln LN7 6DT, UK
Nickerson-Zwaan bv., Postbus 19, 2990-AA Barendrecht, The Netherlands

**Nikon Inc.**, 1300 Walt Whitman Road, Melville, NY 11747-3064, USA
Tel: 001 516 5474200
Fax: 001 516 5470299
URL: http://www.nikonusa.com/

Nikon Corp., Fuji Building, 2–3, 3-chome, Marunouchi, Chiyoda-ku, Tokyo 100, Japan
Tel: 00813 32145311
Fax: 00813 32015856
URL: http://www.nikon.co.jp/main/index_e.htm/

**Novagen**, 601 Science Drive, Madison, WI 53711, USA
Tel: 608 238 6110
Fax: 608 238 1388
URL: http://www.novagen.com/

**Novocastra Laboratories Ltd**, Balliol Business Park West, Benton Lane, Newcastle upon Tyne NE2 8EW, UK

**Nycomed Amersham Imaging**, Amersham Labs, White Lion Road, Amersham, Buckinghamshire HP7 9LL, UK
Tel: 0800 558822 (or 01494 544000)
Fax: 0800 669933 (or 01494 542266)
URL: http//:www.amersham.co.uk/
Nycomed Amersham, 101 Carnegie Center, Princeton, NJ 08540, USA
Tel: 001 609 5146000
URL: http://www.amersham.co.uk/

**Perkin Elmer Ltd**, Post Office Lane, Beaconsfield, Buckinghamshire HP9 1QA, UK
Tel: 01494 676161
URL: http//:www.perkin-elmer.com/

**Pharmacia**, Davy Avenue, Knowlhill, Milton Keynes, Buckinghamshire MK5 8PH, UK
(also see Amersham Pharmacia Biotech)
Tel: 01908 661101
Fax: 01908 690091
URL: http//www.eu.pnu.com/

**Pierce & Warriner (UK) Ltd**, 44 Upper Northgate Street, Chester, Cheshire CH1 4EF, UK
Tel: 44 1244 382 525
Fax: 44 1244 373 212
Pierce Chemical Co., PO Box 117, 3747 N. Meridian Road, Rockford, IL 61105, USA
Tel: 800 874 3723 or 815 968 0747
Fax: 815 968 7316

**Polysciences Inc.**, Corporate Headquarters, 400 Valley Road, Warrington, PA 18976, USA
Tel: 800 523 2575 Fax: 800 343 3291
URL: http://www.polysciences.com/

**Promega UK Ltd**, Delta House, Chilworth Research Centre, Southampton SO16 7NS, UK
Tel: 0800 378994
Fax: 0800 181037
URL: http://www.promega.com/
Promega Corp., 2800 Woods Hollow Road, Madison, WI 53711–5399, USA
Tel: 001 608 2744330
Fax: 001 608 2772516
URL: http://www.promega.com/

**Qiagen UK Ltd**, Boundary Court, Gatwick Road, Crawley, West Sussex RH10 2AX, UK
Tel: 01293 422911
Fax: 01293 422922
URL: http://www.qiagen.com/
Qiagen Inc., 28159 Avenue Stanford, Valencia, CA 91355, USA
Tel: 001 800 4268157
Fax: 001 800 7182056
URL: http://www.qiagen.com/

**Roche Diagnostics Ltd**, Bell Lane, Lewes, East Sussex BN7 1LG, UK
Tel: 0808 1009998 (or 01273 480044)
Fax: 0808 1001920 (01273 480266)
URL: http://www.roche.com/

Roche Diagnostics Corp., 9115 Hague Road, PO Box 50457, Indianapolis, IN 46256, USA
Tel: 001 317 8452358
Fax: 001 317 5762126
URL: http://www.roche.com/
Roche Diagnostics GmbH, Sandhoferstrasse 116, 68305 Mannheim, Germany
Tel: 0049 621 7594747
Fax: 0049 621 7594002
URL: http://www.roche.com/

**Sanyo Gallenkamp**, Monarch Way, Belton Park, Loughborough LE11 5XG, UK

**Santa Cruz Biotechnology, Inc.**, 2161 Delaware Avenue, Santa Cruz, CA 95060, USA
Tel: 831 457 3800 Fax: 831 457 3801
Santa Cruz Biotechnology, Inc., Ludolf-Krehl-Straße 33, 69120 Heidelberg, Germany

**Schleicher and Schuell Inc.**, Keene, NH 03431A, USA
Tel: 001 603 3572398

**Seikagaku America**, 704 Main Street, Falmouth, MA 02540, USA
Tel: 1 508 540 3444; 1 800 237 4512
Fax: 1 508 540 8680
URL: http://www.seikagaku.com

**Shandon Scientific Ltd**, 93–96 Chadwick Road, Astmoor, Runcorn, Cheshire WA7 1PR, UK
Tel: 01928 566611
URL: http//www.shandon.com/

**Sigma–Aldrich Co. Ltd**, The Old Brickyard, New Road, Gillingham, Dorset SP8 4XT, UK
Tel: 0800 717181 (or 01747 822211)
Fax: 0800 378538 (or 01747 823779)
URL: http://www.sigma-aldrich.com/

Sigma Chemical Co., PO Box 14508, St Louis, MO 63178, USA
Tel: 001 314 7715765
Fax: 001 314 7715757
URL: http://www.sigma-aldrich.com/

**Stratagene Inc.**, 11011 North Torrey Pines Road, La Jolla, CA 92037, USA
Tel: 001 858 5355400
URL: http://www.stratagene.com/
Stratagene Europe, Gebouw California, Hogehilweg 15, 1101-CB Amsterdam Zuidoost, The Netherlands
Tel: 00800 91009100
URL: http://www.stratagene.com/

**United States Biochemical (USB)**, PO Box 22400, Cleveland, OH 44122, USA
Tel: 001 216 4649277

**Upstate Biotechnology Inc.**, 1100 Winter Street, Suite 2300, Waltham, MA 02451, USA

**US Biological**, PO Box 261, Swampscott, MA 01907, USA
Tel: 800 520 3011
Fax: 781 639 1768

**Vector Laboratories**, 30 Ingold Road, Burlingame, CA 94010, USA
Tel: 650 697 3600
Fax: 650 697 0339

**VWR Scientific**, Atlanta Regional Distribution Center, 1050 Satellite Blvd., Suwanee, GA 30024, USA
Tel: 770 485 1000
Fax: 770 232 9881
URL: http://www.vwrsp.com/

**Vysis**, 3100 Woodcreek Drive, Downers Grove, IL 60515–5400, USA
Tel: 800 553 7042
Fax: 630 271 7138

Vysis, Rosedale House, Rosedale Road, Richmond, Surrey TW9 2SZ, UK
Tel: 44 181 332 6932
Fax: 44 181 332 2219

**Whatman BioScience Ltd.**, Granta Park, Abington, Cambridge CB1 6GR, UK

**Zymed Laboratories Inc.**, 458 Carlton Court, South San Francisco, CA 94080, USA

# Index